AF347527

Exact Solutions in Classical Field Theory: Solitons, Black Holes and Boson Stars

Exact Solutions in Classical Field Theory: Solitons, Black Holes and Boson Stars

Editors

Nicolas Boulanger
Andrea Campoleoni

MDPI • Basel • Beijing • Wuhan • Barcelona • Belgrade • Manchester • Tokyo • Cluj • Tianjin

Editors

Nicolas Boulanger	Andrea Campoleoni
Physics of the Universe, Fields	Physics of the Universe, Fields
and Gravitation	and Gravitation
UMONS	UMONS
Mons	Mons
Belgium	Belgium

Editorial Office
MDPI
St. Alban-Anlage 66
4052 Basel, Switzerland

This is a reprint of articles from the Special Issue published online in the open access journal *Symmetry* (ISSN 2073-8994) (available at: www.mdpi.com/journal/symmetry/special_issues/exact_solutions_in_classical_field_theory).

For citation purposes, cite each article independently as indicated on the article page online and as indicated below:

LastName, A.A.; LastName, B.B.; LastName, C.C. Article Title. *Journal Name* **Year**, *Volume Number*, Page Range.

ISBN 978-3-0365-2975-2 (Hbk)
ISBN 978-3-0365-2974-5 (PDF)

Contents

Preface to "Exact Solutions in Classical Field Theory: Solitons, Black Holes and Boson Stars"

Compact objects is the name given to solutions in General Relativity (or alternative models of gravity) that are typically so dense that the curvature of space–time around them has detectable effects. An extreme case exists, which is realised and observed in nature: black holes. There exists at least one alternative to black holes, albeit more exotic, given by compact objects made out of bosonic fields, the simplest example being the boson star made of a complex valued, massive scalar field.

The study of compact objects as dense as black holes and boson stars is interesting in its own right. It is also very important from another perspective: Since these objects create very strong gravitational fields, they are also an ideal testing ground for alternative models of gravity and/or for testing the limits of General Relativity. The latter theory works extremely well in the weak regime, but has not been explored in full detail in the very strong regime. This Special Issue is mainly devoted to exact solutions to Einstein–Yang–Mills-type theories and their various extensions. It is dedicated to Prof. Yves Brihaye on the occasion of his 65th birthday. Prof. Brihaye contributed significantly and is still contributing to this very active topic.

Nicolas Boulanger, Andrea Campoleoni
Editors

Article

Gravitating Bubbles of Gluon Plasma above Deconfinement Temperature

Yves Brihaye [1] and Fabien Buisseret [2,3,*]

[1] Service de Physique de l'Univers, Champs et Gravitation, UMONS Research Institute for Complex Systems, Université de Mons, Place du Parc 20, 7000 Mons, Belgium; yves.brihaye@umons.ac.be

[2] Service de Physique Nucléaire et Subnucléaire, UMONS Research Institute for Complex Systems, Université de Mons, Place du Parc 20, 7000 Mons, Belgium

[3] CeREF, Chaussée de Binche 159, 7000 Mons, Belgium

* Correspondence: fabien.buisseret@umons.ac.be

Received: 2 September 2020; Accepted: 8 October 2020; Published: 13 October 2020

Abstract: The equation of state of SU(3) Yang–Mills theory can be modelled by an effective Z_3-symmetric potential depending on the temperature and on a complex scalar field ϕ. Allowing ϕ to be dynamical opens the way to the study of spatially localized classical configurations of the scalar field. We first show that spherically symmetric static Q-balls exist in the range $(1 - 1.21) \times T_c$, T_c being the deconfinement temperature. Then we argue that Q-holes solutions, if any, are unphysical within our framework. Finally, we couple our matter Lagrangian to Einstein gravity and show that spherically symmetric static boson stars exist in the same range of temperature. The Q-ball and boson-star solutions we find can be interpreted as "bubbles" of deconfined gluonic matter; their mean radius is always smaller than 10 fm.

Keywords: deconfinement; Matter-gravity coupling; Yang–Mills theory; Q-ball; boson star

---

1. Introduction

A fascinating feature of Yang–Mills theory is the existence of a deconfinement temperature, T_c, above which free color charges (free gluons) may propagate without being confined into color singlets [1,2]. This deconfined phase can be thought as a "gluon plasma", in analogy with the celebrated quark-gluon plasma experimentally created first at RHIC [3,4], i.e., the deconfined phase of full QCD.

In pure SU(N_c) Yang–Mills theory, deconfinement might be driven by the breaking of a global center symmetry, i.e., Z_{N_c} symmetry [5,6], exemplified by the behavior of the Polyakov loop at finite temperature T. The Polyakov loop is defined as $L(T, \vec{y}) = \frac{1}{N_c} \mathrm{Tr}_c P\, e^{ig \int_0^{1/T} d\tau A_0(\tau, \vec{y})}$, with A_0 the temporal component of the Yang–Mills field and $\vec{y}$ the spatial coordinates. P is the path-ordering, g is the strong coupling constant and units where $\hbar = c = k_B = 1$ are used. Since gauge transformations belonging to the center of the gauge group only cause $L(T, \vec{y})$ to be multiplied by an overall factor, the Polyakov loop is such that the spatial average $\langle L \rangle = 0 \ (\neq 0)$ when the Z_{N_c} symmetry is present (broken), hence when the theory is in a (de)confined phase [2,7]. Hence $\langle L \rangle$ is commonly seen as an order parameter for Yang–Mills theory at finite T, although it may not be the best physical candidate. As shown in [8], distinct Z_{N_c} phases above T_c do not actually label different physical states. Moreover, the value $-T \ln \langle L \rangle$, giving the free energy of a static color source in the heath bath, may take unphysical complex values above T_c. The real number $|\langle L \rangle|$ may be a better candidate, with a free energy given by $-T \ln |\langle L \rangle|$. It is also argued in [8] that the correlator $\left\langle L(T, \vec{y}) L(T, \vec{0})^\dagger \right\rangle$ at large distances may be a proper order parameter. Another relevant order parameter for pure Yang–Mills theory has finally been formulated in [9]: It is the spatial 't Hooft loop $V(C)$ [10], C being a closed spatial contour. $V(C)$ shows a perimeter law in the confined phase and an area law in the deconfined phase: $V(C) \sim \exp(-m\, P(C))$ and $V(C) \sim \exp(-\alpha\, S(C))$ respectively, with P and S the perimeter and the area of the closed contour.

In view of the above results we can safely assume that there exists one scalar field ϕ playing the role of an order parameter. Moreover, there should exist a Z_{N_c} temperature-dependent potential $V(\phi, \phi^*, T)$, whose global minimum is different at $T < T_c$ and $T > T_c$, reproducing the equation of state computed in lattice QCD [11] in the mean-field approximation—recall that the pressure reads $p = -\min_{\phi,\phi^*} V$. In the case $N_c = 3$, the Z_3-symmetry should be present through terms in $\phi^3 + \phi^{*3}$ at the lowest-order in a power expansion of V [12]: The explicit form we will use is given in Section 2. In the following we go beyond mean-field theory and treat ϕ as a complex, position-dependent, scalar field: This dynamical field mimics the behavior of pure Yang–Mills theory at finite T. The value of $|\phi|$ being related to the phase of the gluonic matter, we can summarize the aim of our study as follows: We search for configurations describing localized regions of (de)confined gluonic matter, either in flat or curved 750/9.

On one hand we have already shown the existence of nontrivial solutions for $\phi(\vec{y})$, vanishing at infinity, at the deconfinement temperature in flat space-time and at large N_c [13]. On the other hand, nontrivial static configurations in Z_3-symmetric potentials have already been found in [14,15]. Most of the effort in the field has been devoted to study the temporal evolution of such solutions in close relation with thermalization issues of experimentally observed quark-gluon-plasma [14,16–19]. In this work we search for spherically symmetric static Q-ball solutions with a focus on conditions constraining their existence: temperature range, radial nodes, etc. Less standard solitons as Q-holes [20], never studied up to now within that framework, are also discussed. Q-balls and Q-holes are discussed in Sections 3 and 5.

Finally, we couple our Z_3-symmetric Lagrangian to Einstein gravity. To our knowledge, very few attempts to describe to interplay between gravity and confinement/deconfinement phase transition can be found in the literature. One can quote [21,22], respectively discussing the loss of simultaneity between chiral restoration and deconfinement in curved space, and the possible existence of deconfined regions near a black hole horizon. Here we go one step further by building "particle-like" solutions for our scalar field that are known to appear in pure $3 + 1$-dimensional Yang–Mills theory coupled to Einstein gravity, see the seminal paper [23]. Within our approach the Yang–Mills degrees of freedom are replaced by a complex scalar field, whose associated Q-balls, when coupled to gravity, are called boson stars—see the review [24] for more recent references. To our knowledge, such a problem has never been addressed at finite temperature although research devoted to "QCD boson stars" (at $T = 0$) is currently ongoing [25]. We build gravitating solutions of static-boson-star-type, i.e., spherically symmetric localized configurations of the scalar field that lead to an asymptotically flat metric without singularity, see Section 5.

2. The Model

2.1. Z_3-Symmetric Potential

Let us model SU(3) Yang–Mills theory at finite temperature by an effective Lagrangian based on a complex scalar field ϕ plus Z_3−symmetry. We use the potential V of Ref. [26] which reads

$$U(\phi, \phi^*, T) = \frac{V(\phi, \phi^*, T)}{T^4}, \tag{1}$$

with

$$U(\phi, \phi^*, T) = -\frac{b_2(T)}{2}|\phi|^2 + b_4(T) \ln \left[1 - 6|\phi|^2 + 4(\phi^3 + \phi^{*3}) - 3|\phi|^4 \right], \tag{2}$$

and

$$b_2(T) = 3.51 - 2.47 \frac{T_c}{T} + 15.22 \left(\frac{T_c}{T} \right)^2, \quad b_4(T) = -1.75 \left(\frac{T_c}{T} \right)^3. \tag{3}$$

We have retained the above parametrization because it to an optimal agreement with the equation of state of pure SU(3) Yang–Mills theory computed in lattice QCD [11], and also with the full

$N_f = 2$ lattice QCD equation of state at zero and nonzero chemical potential when coupled to a Nambu–Jona–Lasinio (NJL) model [26]. We note that the latter reference explicitly identifies the scalar field with the Polyakov loop. In fact, their result is more general since ϕ can be regarded as the actual order parameter of the model, not necessarily the Polyakov loop.

Potential (1) is displayed in Figure 1 for the values (3) of the parameters and for several temperatures. The change in minimum is clearly seen above and below T_c: A (non)vanishing value for $|\phi|$ gives the minimum of U in the (de)confined phase. We notice that potential (1) is only Z_3-symmetric and not U(1)-symmetric as is often the case in Lagrangians based on a complex scalar field, with typical potentials of the form $|\phi|^6 - 2|\phi|^4 + b|\phi|^2$ [27]. A U(1)-symmetry can be recovered in the large-N_c limit of Z_{N_c}-symmetric potentials, see [13,28].

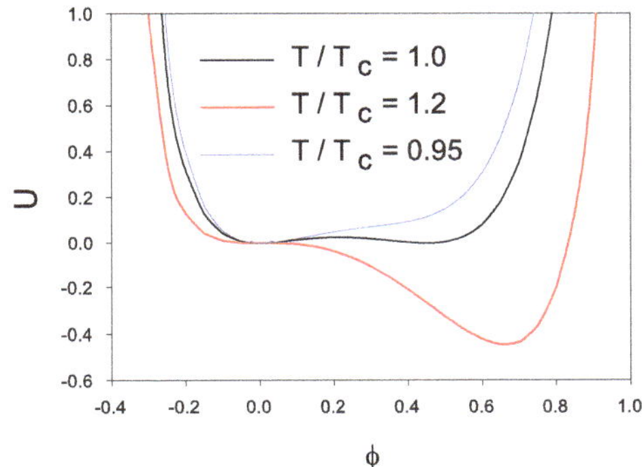

Figure 1. The potential $U(\phi, \phi^*, T)$ versus ϕ for various temperatures. U is given by Equations (2) and (3), and the plot is restricted to $\phi \in \mathbb{R}$ for the sake of clarity.

According to the suggestion of e.g., Ref. [29], we choose for ϕ a kinetic part of the form $N_c^2 T^2 \partial_\mu \phi \partial^\mu \phi^* / \lambda$, which has both the correct energy dimensions and the expected N_c-scaling when the gauge group SU(N_c) is chosen. λ is the 't Hooft coupling. Minkowski metric has signature $(+ - - -)$. In our SU(3) case, recalling that $\alpha_s = \lambda / (12\pi)$, we can write our Lagrangian as

$$\mathcal{L}_{phys} = \frac{3T^2}{4\pi\alpha_s} \partial_\mu \phi \partial^\mu \phi^* - T^4 U(\phi, \phi^*, T), \tag{4}$$

where $\phi = \phi(y^\mu)$, y^μ are the space-time coordinates. It is convenient to further define dimensionless variables x^μ related to the original (physical) ones by

$$y^\mu = l_{phys}\, x^\mu, \quad \text{with} \quad l_{phys} = \frac{\sqrt{3}}{T\sqrt{4\pi\alpha_s}}, \tag{5}$$

so that the above Lagrangian can be replaced by the dimensionless one

$$\mathcal{L} = \frac{\mathcal{L}_{phys}}{T^4} = \partial_\mu \phi \partial^\mu \phi^* - U(\phi, \phi^*, T), \tag{6}$$

where $\phi = \phi(x^\mu)$ and where T is expressed in units of T_c.

It is worth estimating the physical length used in the model. First, a typical value for the deconfinement temperature in pure gauge QCD is $T_c = 0.3$ GeV [30]. Second, a way to estimate α_s is to note that the short-range part of the static interaction between a quark and an antiquark scales as $-(4/3)\alpha_s/r$, at least from $T = 0$ to T_c. Lattice studies, performed at $N_c = 3$, favor $\alpha_s = 0.2$ up to $T = T_c$ [31], which is the value we retain here. We are then in position to estimate that, at $T = T_c$,

$$l_{phys} = 3.6 \, \text{GeV}^{-1} = 0.72 \, \text{fm}. \tag{7}$$

2.2. Coupling to Einstein Gravity

The coupling of the above Lagrangian to gravity can be performed by minimally coupling the scalar field to Einstein gravity: The action reads

$$S = \int d^4x \sqrt{-g} \left(\frac{R}{\alpha} + \mathcal{L} \right), \tag{8}$$

with the effective coupling constant

$$\alpha = 16\pi G_N l_{phys}^2 T^4. \tag{9}$$

The replacement of the partial derivatives by covariant ones in (6) must be performed.

3. Q-Balls

3.1. Ansatz and Existence Conditions

We begin by considering Lagrangian (6) where ϕ^3 and ϕ^{*3} are replaced by $|\phi|^3$ in order to recover the usual U(1)-symmetry needed to build Q-balls solutions.

The classical equations of motion in flat space-time with potential $U(|\phi|, T) = -\frac{b_2(T)}{2}|\phi|^2 + b_4(T)\ln\left[1 - 6|\phi|^2 + 8|\phi|^3 - 3|\phi|^4\right]$ read

$$\partial_\mu \partial^\mu \phi = \partial_{\phi^*} U = -\frac{b_2(T)}{2}\phi + 6b_4(T)\frac{-\phi + 2|\phi|\phi - |\phi|^2\phi}{1 - 6|\phi|^2 - 3|\phi|^4 + 8|\phi|^3} \tag{10}$$

plus the complex conjugated equation. We then perform the usual Q-ball ansatz on the scalar field :

$$\phi = \exp(i\omega t)\phi(r), \tag{11}$$

where $t = x^0$ and where $r = \sqrt{(x^1)^2 + (x^2)^2 + (x^3)^2}$. The solutions we will build can be characterized by their mas M and by a dimensionless conserved charge Q, respectively defined by

$$M = M_{phys}\int d^3x \, T_{00} \tag{12}$$

with $M_{phys} = 1/l_{phys}$ and

$$Q = 2\omega \int d^3x \, |\phi|^2. \tag{13}$$

The temporal component of the energy-momentum tensor represents the energy density, given by

$$T_{00} = \omega^2|\phi|^2 + \vec{\nabla}\phi \cdot \vec{\nabla}\phi^* + U(|\phi|). \tag{14}$$

The conserved charge Q finds its origin in the (artificially restored) U(1)-symmetry of the considered Lagrangian, leading to a conserved Noether current of the form $J_\mu = i(\phi\partial_\mu\phi^* - \phi^*\partial_\mu\phi)$, Q being the space integral of J_0. Axially symmetric solutions with $k \neq 0$ are spinning Q-balls whose angular momentum J is related to the charge Q according to $J = kQ$ [27]. Here we focus on non-spinning Q-balls.

We have studied the equations for generic values of ω although it is clear that only the solutions corresponding to $\omega = 0$ are physically relevant for the potential under consideration: The original potential is Z_3-symmetric, not U(1). Note also that if $\phi(r)$ is a real solution of the equations of motion, $e^{\frac{ik\pi}{3}}$ with $k \in \mathbb{Z}$ is also a solution because of the system's symmetry.

The mass term of the potential plays a crucial role in the existence of the solutions. In a power expansion in $|\phi|$,

$$U(|\phi|, T) = m^2(T)|\phi|^2 + \text{'higher order''} \ , \ \text{with} \ m^2(T) = -\frac{b_2(T)}{2} - 6b_4(T). \tag{15}$$

General results on Q-balls [27] state that the soliton exist for $\omega_{min} \leq \omega \leq \omega_{max}$ with

$$\omega_{min} = \min_{|\phi|} \frac{U(|\phi|, T)}{|\phi|^2} \ , \ \omega_{max} = m(T). \tag{16}$$

In particular, if the potential $U(|\phi|)$ is negative in some interval of values of $|\phi|$, the value $\omega = 0$ belongs to the spectrum of the boson star. This turns out to be the case for $T > T_c$. The condition $m(T)^2 > 0$ also needs to be fulfilled; in terms temperature, this corresponds to $T/T_c > 1.21$. As a consequence, the general properties of Q-balls solutions suggest that Q-ball solutions with zero frecuency will exist for $1 < T/T_c < 1.21$.

The parametrization (2) of the potential is not unique. In particular, a power expansion of the form

$$U = -\frac{b_2}{2}|\phi|^2 - \frac{b_3}{6}(\phi^3 + \phi^{*3}) + \frac{b_4}{4}|\phi|^4 \tag{17}$$

is often used in effective YM theories at finite T, the choice $b_2 = 6.75 - 1.95(T_c/T) + 2.63(T_c/T)^2 - 7.44(T_c/T)^3$, $b_3 = 0.75$, $b_4 = 7.5$ leading to a good agreement with lattice QCD data [32]. Using this alternative choice would not forbid the existence of Q-ball solutions: The mass term b_2 is positive above T_c and the criterion (16) leads to $\omega_{min} < 0$, so $\omega = 0$ solitons are allowed also in this case.

3.2. Numerical Results

A numerical resolution of the equations (10) can now be performed. We use a collocation method for boundary-value ordinary differential equations, equipped with an adaptive mesh selection procedure [33]. The regularity of the solution at the origin implies $\frac{d\phi}{dr}(r = 0) = 0$, the finiteness of the energy and the charge impose $\phi(\infty) = 0$. These are the boundary conditions.

We present in Figure 2 the spectrum of the Q-balls for $T/T_c = 0.83, 1.01, 1.18$. It can be observed that no Q-ball solution with $\omega = 0$ can be found below T_c: It is a nice feature of our model that it does not lead to solutions modelling deconfined matter below T_c. In the range $1 < T/T_c < 1.21$ suggested by the above analysis however, such solutions can be found. From now on, we concentrate on the latter $\omega = 0$ solutions. Our results are summarized in Figures 3 and 4.

Our numerical analysis shows that in the limit $T/T_c \to 1.21$ the scalar function $\phi(r)$ approaches uniformly the null function as expected by the existence criterion discussed before. The limit $T \to T_c$ reveals a peculiar behavior of the solitons: Their mean radius and mass increase considerably. In this limit the scalar function $\phi(r)$ is closer and closer to a nonzero constant solution, leading to the observed increase in mass and mean radius.

All the Q-balls we find have a mean radius smaller than $14 \times l_{phys} = 10$ fm and are lighter than $140 \times M_{phys} = 38.9$ GeV. Similar solutions were found in [14] with a simpler, power-law, Z_3-symmetric potential of the form $|\phi|^2 - a(\phi^3 + \phi^{*3}) + b|\phi|^4$. At $T = 1.1\,T_c$ they find a soliton with a typical size of $1.5-2$ fm while we find a Q-ball with mean radius 1.43 fm and mass 200 MeV at the same temperature. Other results are obtained in [14] but in $2 + 1$ dimensions so they cannot be compared to ours.

We have tried to construct radially excited solutions, i.e., solutions where the radial function presents one or more nodes, but so far, we cannot find any. The absence of solutions presenting nodes for our model can be explained by the following argument. In general, the existence of node solutions is closely related to the shape of the effective potential

$$V_{eff}(|\phi|) = \frac{\omega^2}{2}|\phi|^2 - \frac{1}{2}U(|\phi|) \,. \tag{18}$$

Several conditions are necessary for node solutions to exist [27,34]: (i) $\phi = 0$ should be a local maximum of V_{eff}, (ii) the effective potential should admit local minima for both signs of ϕ. It would be challenging to have a generic proof of the absence of node solutions with our potential but an inspection of the potential $U(|\phi|)$ quickly reveals that no local minimum exist for $\phi < 0$ when $\phi \in \mathbb{R}$ (see Figure 1), so the condition (ii) cannot be fulfilled when $\omega = 0$.

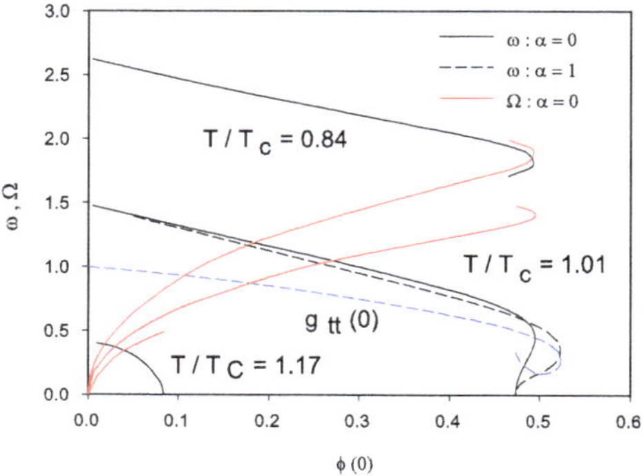

Figure 2. Relation between ω and $\Omega \equiv \sqrt{m^2(T) - \omega^2}$ versus $\phi(0)$ for three values of T/T_c in flat space-time ($\alpha = 0$) (solid lines). The dotted lines represent ω and $g_{tt}(0)$ in the case $T/T_c = 1.01$ for gravitating solutions ($\alpha = 1$).

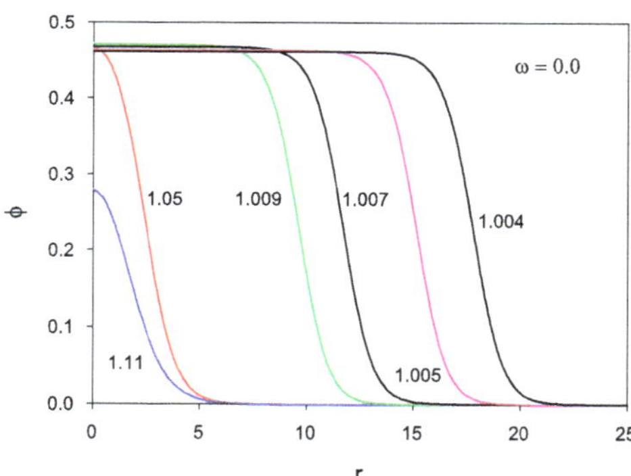

Figure 3. Profiles of $\phi(r)$ for several values of T/T_c. Distances are in units of l_{phys}.

The spectrum of the fundamental Q-ball appears quite different with the present potential than in more conventional U(1)-symmetric potentials. For example, one of us previously studied Q-balls with the SUSY-inspired potential $U_{SUSY}(|\phi|) \sim 1 - \exp\left(-\frac{|\phi|^2}{\eta^2}\right)$, with $\eta \in \mathbb{R}_0^+$ [35]. In contrast to our potential, solutions can be constructed for arbitrarily large values of the central density $\phi(0)$ with the latter potential, and radially excited solitons can be obtained. We considered an effective potential consisting of a linear superposition of our potential and the SUSY-potential. The latter is known to admit node solutions: $U_{eff} = \cos(y)\, U_{SUSY} + \sin(y)\, U$ with $y \in [0, \pi/2]$. It turns out that when we progressively deform the SUSY-potential into our potential (say with a fixed value $\phi(0) < 1$) the zero-node solutions get continuously deformed and the frequency ω decreases with increasing the mixing parameter y. By contrast, for the one node solution, the ω quickly reaches $\omega = 1$ and the solution becomes oscillating.

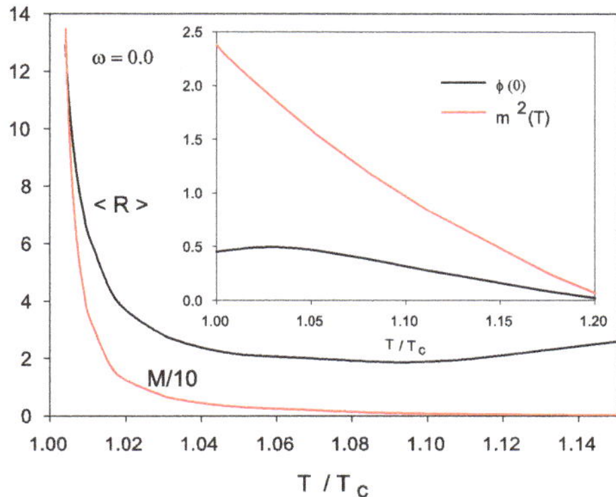

Figure 4. Evolution of the mean radius $\langle R \rangle$ as a function of T/T_c, the insert contains $\phi(0)$ and the mass (12) of the scalar field. Distances are in units of l_{phys} and masses are in units of M_{phys}.

3.3. Symmetry Breaking

An obvious outlook is to include the matter sector of QCD in our approach. As discussed in [12], an immediate effect of quarks is the breaking of Z_3-symmetry in the potential and the simplest way to mimic that symmetry breaking is to add a term proportional to $(\phi + \phi^*)$ or even $(\phi + \phi^*)^2$ to the potential. A more rigorous treatment of quark fields may be achieved by coupling ϕ to a NJL Lagrangian [15,36]. Within mean-field approximation for quarks, it has been show in [37] that complex-valued solutions of Q-ball-type still exist with broken Z_3-symmetry. A question arising at this stage is therefore: May the $\omega = 0$ Q-balls we constructed "survive" to such a symmetry breaking?

We propose to perform the substitution $U \rightarrow U_\beta = U + \beta\,(\phi + \phi^*)^2$ with β a real constant parameter. This ansatz breaks the Z_3-symmetry while still allowing analytical calculations. In a power expansion in ϕ on the real axis, the β-term shifts the mass term: $m^2(T) \rightarrow m^2(T) + 4\beta$. Graphical inspection of the modified potential shows that there is always an interval of ϕ values in which U_β is negative above T_c if $\beta < 0$. Nevertheless a negative value of β lowers m^2 and therefore lowers the maximal temperature at which Q-balls may be expected (m^2 is indeed a decreasing function of T). If $-0.5 < \beta < 0$, there always exists an interval of temperatures above T_c for which the existence of Q-balls is guaranteed. We can thus safely assume that the solutions we find will not necessarily disappear in a more realistic theory including quarks.

4. Q-Holes

Q-holes are configurations of the scalar field behaving like (11) but such that $|\phi(r = 0)| > 0$ and $|\phi(r \rightarrow \infty)| = \phi_c > |\phi(r = 0)|$, with ϕ_c a local minimum of the potential under study [20]. Are $\omega = 0$ Q-holes solutions worth being constructed within the present approach?

A necessary condition for Q-holes to exist is that the second maximum of the effective potential (18) is lower than the maximum at the origin [38]. As illustrated in Figure 1, it can only happen below T_c in our model since $V_{eff}(|\phi|) = -\frac{1}{2}U(|\phi|)$ at vanishing ω. A configuration where ϕ is everywhere nonzero at a temperature below the deconfinement one is not physically relevant; although the problem is interesting from a technical point of view we have thus to discard Q-hole solutions in the present work. The same conclusion holds for the parameterization (17).

5. Boson Stars

The action (8) leads to the Einstein equation

$$G_{\mu\nu} = \frac{\alpha}{2} T_{\mu\nu} \tag{19}$$

where the energy-momentum tensor is given by $T_{\mu\nu} = (D_\mu\phi)^*(D_\nu\phi) + (D_\mu\phi)(D_\nu\phi)^* - g_{\mu\nu}(D_\alpha\phi)^*(D^\alpha\phi) + g_{\mu\nu}\mathcal{L}$. The metric defines $ds^2 = g_{\mu\nu}dx^\mu dx^\nu$ and D_μ is the covariant derivative. An estimation of this coupling constant α in the temperature range under study is $16\pi G_N l^2_{phys} T_c^4 = 3.58 \times 10^{-38}$. It is so small that no significant change of the solutions can be observed by numerical investigation. To appreciate more clearly the influence of gravity on the system we will construct solutions with $\alpha = 0.01$ and 1.

We search for boson-star solution. First, the ansatz (11) and the boundary conditions are kept for the scalar field. Second, we choose a spherically symmetric ansatz for the metric:

$$ds^2 = -f(r)dt^2 + \frac{l(r)}{f(r)}\left(dr^2 + r^2\, d\theta^2 + r^2\sin^2\varphi\, d\varphi^2\right), \tag{20}$$

with the boundary conditions $f(r = +\infty) = l(r = +\infty) = 1$ (asymptotically flat space-time) and $f'(r = 0) = l'(r = 0) = 0$ (no singularity at origin). The gravitational mass M_G of these gravitating objects is defined as usual according to $f(r \to \infty) \sim 1 - \frac{2M_G G_N}{r}$. The explicit equations involving ϕ, f and l can be found in Appendix B of [34]; we do not recall them here for the sake of simplicity. The same numerical method as for Q-balls is used to construct boson-star solutions [33].

Even for such large values as $\alpha = 1$, our results indicate that gravitating solutions with $\omega = 0$ still exist on roughly the same interval of T/T_c, see Figure 5. However, the numerical analysis turns out to be tricky in the limit $T \to T_c$ likely because the local minimum of the potential disappears. Our results strongly suggest that the gravitational mass and mean radius increase considerably in the limit $T \to T_c$ as shown by Figure 5, similarly as what is observed in the Q-ball case. As expected, the metric gets more deviated from the Minkowski metric in the central region of the soliton: for instance $g_{00} \ll 1$ (see the blue line of Figure 5) and one can expect an essential singularity of the metric to be formed at T_c.

The profiles of the solution corresponding to $T/T_c = 1.01$ are presented in Figure 6 for $\alpha = 1$ (solid lines). This plot clearly demonstrates that the soliton splits the space into two distinct regions: an interior region where ϕ is practically constant and strongly curving space-time and a region with $\phi \sim 0$ where space-time is essentially Minkowski. These regions are separated by a "wall" of the scalar field. The profile of a solution at an intermediate temperature $T/T_c = 1.11$ is also shown in Figure 6; the same qualitative features are observed. The boson star finally presents different features for $T/T_c \to 1.21$, i.e., the limit of vanishing $m(T)$. In this limit the scalar field approaches uniformly the null function and the Minkowski space-time is approached.

The existence of boson-star configurations for $\alpha = 1$ implies the existence of such solutions for much smaller, "realistic", values of the coupling constant around T_c, see Figure 5. Since potential (17) will also lead to Q-balls above T_c, we can state that boson stars also exist with that parameterization. We choose however not to perform full numerical computations since potential (2) leads to a more accurate modelling of QCD equation of state as computed on the lattice.

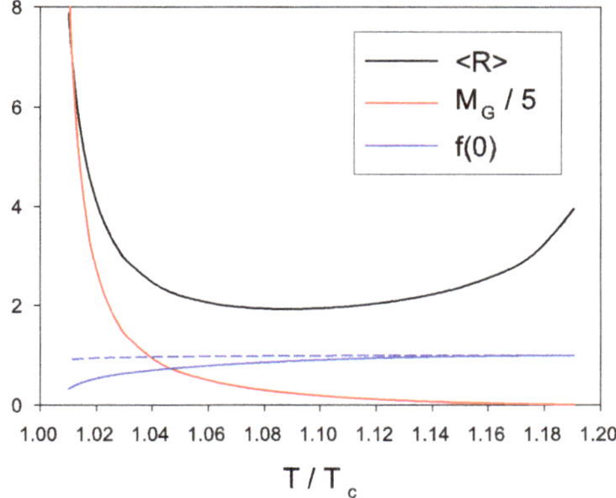

Figure 5. Evolution of the mean radius $< R >$ (black line, in units of l_{phys}) and of the gravitational mass M_G (red lines, in units of M_{phys}) as function of T_c/T for $\omega = 0$ boson stars. The metric component $g_{00} = f(0)$ is represented by the solid (resp. dashed) blue lines for $\alpha = 1$ (resp. $\alpha = 0.1$). These values depend very weakly on α and the curves are mostly superimposed.

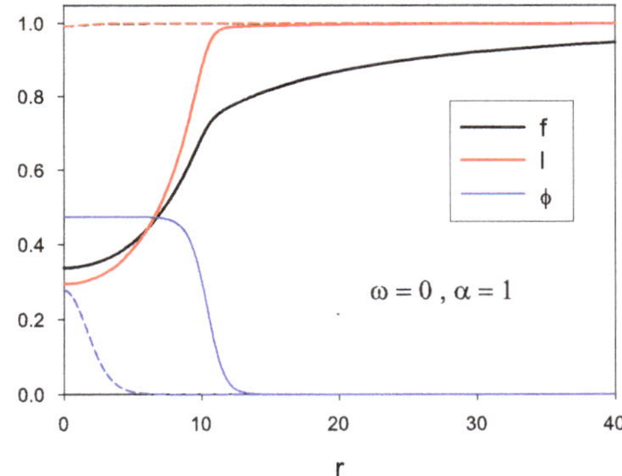

Figure 6. Profiles of the metric functions f, l and of the scalar field ϕ for $\alpha = 1, \omega = 0$ and two values of the temperature : $T/T_c = 1.01$ (solid lines) $T/T_c = 1.11$ (dashed lines). The radial variable r is in units of l_{phys}.

6. Summary and Outlook

We have built Q-balls and boson stars from a model with a complex scalar field plus a temperature-dependent Z_3-symmetric potential mimicking Yang–Mills theory at finite temperature. We have shown that static Q-balls only exist between 1 and 1.21 T_c with a mean radius smaller than 10 fm and that they cannot have radial nodes. The solutions we find are spherically symmetric and the scalar field is such that $|\phi(r = 0)| \neq 0$ and $|\phi(r \to \infty)| = 0$; they can be interpreted as "bubbles" of deconfined gluonic matter. We also showed that Q-holes solutions should be discarded from a physical point of view since they are solutions modelling a deconfined phase, but that can only exist below T_c within our approach. Static boson stars exist in roughly the same temperature range as Q-balls. Their qualitative features are almost independent on the value of the matter-Einstein gravity coupling constant α.

To our knowledge, it is the first time that boson stars are constructed from an effective potential such as (2). Typical potentials used in boson-star-related studies are such that solutions exist for $0 < \omega_{min} \leq \omega \leq \omega_{max}$, see i.e., [35]. It is worth pointing out that the potential used here even allows the existence of static solutions with $\omega_{min} = 0$.

Computation of the QCD equation of state in curved space-time shows that the latter may affect the phase-diagram of the theory by increasing the splitting between the critical points for chiral and deconfinement transitions [21]. We hope to present generalizations of our boson-star configurations to the case of a nontrivial quark field in a future work; they could shed new light on the interplay between confinement, chiral symmetry and gravity.

Author Contributions: Conceptualization, F.B. and Y.B.; software, Y.B.; validation, F.B. and Y.B.; formal analysis, F.B. and Y.B.; writing—original draft preparation, F.B. and Y.B.; writing—review and editing, F.B. and Y.B. All authors have read and agreed to the published version of the manuscript.

Funding: This research received no external funding.

Conflicts of Interest: The authors declare no conflict of interest.

References

1. Polyakov, A.M. Thermal Properties of Gauge Fields and Quark Liberation. *Phys. Lett. B* **1978**, *72*, 477–480. doi:10.1016/0370-2693(78)90737-2.
2. Susskind, L. Lattice Models of Quark Confinement at High Temperature. *Phys. Rev. D* **1979**, *20*, 2610–2618. doi:10.1103/PhysRevD.20.2610.
3. Arsene, I.; Bearden, I.G.; Beavis, D.; Besliu, C.; Budick, B.; Bøggild, H.; Chasman, C.; Christensen, C.H.; Christiansen, P.; Cibor, J.; et al. Quark gluon plasma and color glass condensate at RHIC? The Perspective from the BRAHMS experiment. *Nucl. Phys. A* **2005**, *757*, 1–27. doi:10.1016/j.nuclphysa.2005.02.130.
4. Aidala, C.; Akiba, Y.; Alfred, M.; Andrieux, V.; Aoki, K.; Apadula, N.; Asano, H.; Ayuso, C.; Azmoun, B.; Babintsev, V.; et al. Creation of quark–gluon plasma droplets with three distinct geometries. *Nat. Phys.* **2019**, *15*, 214–220. doi:10.1038/s41567-018-0360-0.
5. Yaffe, L.G.; Svetitsky, B. First Order Phase Transition in the SU(3) Gauge Theory at Finite Temperature. *Phys. Rev. D* **1982**, *26*, 963. doi:10.1103/PhysRevD.26.963.
6. Svetitsky, B.; Yaffe, L.G. Critical Behavior at Finite Temperature Confinement Transitions. *Nucl. Phys. B* **1982**, *210*, 423–447. doi:10.1016/0550-3213(82)90172-9.
7. Weiss, N. The Wilson Line in Finite Temperature Gauge Theories. *Phys. Rev. D* **1982**, *25*, 2667. doi:10.1103/PhysRevD.25.2667.
8. Smilga, A.V. Are Z(N) bubbles really there? *Ann. Phys.* **1994**, *234*, 1–59. doi:10.1006/aphy.1994.1073.
9. Korthals-Altes, C.; Kovner, A.; Stephanov, M.A. Spatial't Hooft loop, hot QCD and Z(N) domain walls. *Phys. Lett. B* **1999**, *469*, 205–212, doi:10.1016/S0370-2693(99)01242-3.
10. Hooft, G.T. On the Phase Transition Towards Permanent Quark Confinement. *Nucl. Phys. B* **1978**, *138*, 1–25. doi:10.1016/0550-3213(78)90153-0.
11. Boyd, G.; Engels, J.; Karsch, F.; Laermann, E.; Legeland, C.; Lutgemeier, M.; Petersson, B. Thermodynamics of SU(3) lattice gauge theory. *Nucl. Phys. B* **1996**, *469*, 419–444. doi:10.1016/0550-3213(96)00170-8.
12. Sannino, F. Higher representations: Confinement and large N. *Phys. Rev. D* **2005**, *72*, 125006. doi:10.1103/PhysRevD.72.125006.
13. Brihaye, Y.; Buisseret, F. Q-ball formation at the deconfinement temperature in large-N_c QCD. *Phys. Rev. D* **2013**, *87*, 014020. doi:10.1103/PhysRevD.87.014020.
14. Gupta, U.S.; Mohapatra, R.K.; Srivastava, A.M.; Tiwari, V.K. Simulation of Z(3) walls and string production via bubble nucleation in a quark-hadron transition. *Phys. Rev. D* **2010**, *82*, 074020. doi:10.1103/PhysRevD.82.074020.
15. Jin, J.; Mao, H. Nontopological Soliton in the Polyakov Quark Meson Model. *Phys. Rev. C* **2016**, *93*, 015202. doi:10.1103/PhysRevC.93.015202.
16. Scavenius, O.; Dumitru, A.; Jackson, A.D. Explosive decomposition in ultrarelativistic heavy ion collision. *Phys. Rev. Lett.* **2001**, *87*, 182302. doi:10.1103/PhysRevLett.87.182302.

17. Fraga, E.S.; Krein, G. Can dissipation prevent explosive decomposition in high-energy heavy ion collisions? *Phys. Lett. B* **2005**, *614*, 181–186. doi:10.1016/j.physletb.2005.03.079.

18. Gupta, U.S.; Mohapatra, R.K.; Srivastava, A.M.; Tiwari, V.K. Effects of Quarks on the Formation and Evolution of Z(3) Walls and Strings in Relativistic Heavy-Ion Collisions. *Phys. Rev. D* **2012**, *86*, 125016. doi:10.1103/PhysRevD.86.125016.

19. Mohapatra, R.K.; Srivastava, A.M. Domain growth and fluctuations during quenched transition to quark-gluon plasma in relativistic heavy-ion collisions. *Phys. Rev. C* **2013**, *88*, 044901. doi:10.1103/PhysRevC.88.044901.

20. Nugaev, E.; Shkerin, A.; Smolyakov, M. Q-holes. *J. High Energy Phys.* **2016**, *12*, 32. doi:10.1007/JHEP12(2016)032.

21. Sasagawa, S.; Tanaka, H. The separation of the chiral and deconfinement phase transitions in the curved space-time. *Prog. Theor. Phys.* **2012**, *128*, 925–939. doi:10.1143/PTP.128.925.

22. Flachi, A. Deconfinement transition and Black Holes. *Phys. Rev. D* **2013**, *88*, 041501. doi:10.1103/PhysRevD.88.041501.

23. Bartnik, R.; Mckinnon, J. Particle-Like Solutions of the Einstein Yang-Mills Equations. *Phys. Rev. Lett.* **1988**, *61*, 141–144. doi:10.1103/PhysRevLett.61.141.

24. Liebling, S.L.; Palenzuela, C. Dynamical Boson Stars. *Living Rev. Relativ.* **2012**, *15*, 6. doi:10.12942/lrr-2012-6, 10.1007/s41114-017-0007-y.

25. Kulshreshtha, U.; Kumar, S.; Kulshreshtha, D.S.; Kunz, J. Boson Stars and QCD Boson Stars. *PoS* **2020**, *LC2019*, 054, [arXiv:hep-th/2001.03745]. doi:10.22323/1.374.0054.

26. Ratti, C.; Roessner, S.; Thaler, M.A.; Weise, W. Thermodynamics of the PNJL model. *Eur. Phys. J. C* **2007**, *49*, 213–217, doi:10.1140/epjc/s10052-006-0065-x.

27. Volkov, M.S.; Wohnert, E. Spinning Q balls. *Phys. Rev. D* **2002**, *66*, 085003, doi:10.1103/PhysRevD.66.085003.

28. Buisseret, F.; Lacroix, G. A large-N_c PNJL model with explicit Z_{N_c} symmetry. *Phys. Rev. D* **2012**, *85*, 016009, doi:10.1103/PhysRevD.85.016009.

29. Dumitru, A.; Pisarski, R.D. Event-by-event fluctuations from decay of a Polyakov loop condensate. *Phys. Lett. B* **2001**, *504*, 282–290, doi:10.1016/S0370-2693(01)00286-6.

30. Sarkar, S.; Satz, H.; Sinha, B. *The Physics of the Quark-Gluon Plasma*; Lecture Notes in Physics; Springer: Berlin/Heidelberg, Germany, 2010; Volume 785, pp. 1–369. doi:10.1007/978-3-642-02286-9.

31. Kaczmarek, O.; Zantow, F. Static quark anti-quark interactions in zero and finite temperature QCD. I. Heavy quark free energies, running coupling and quarkonium binding. *Phys. Rev. D* **2005**, *71*, 114510. doi:10.1103/PhysRevD.71.114510.

32. Ratti, C.; Thaler, M.A.; Weise, W. Phases of QCD: Lattice thermodynamics and a field theoretical model. *Phys. Rev. D* **2006**, *73*, 014019. doi:10.1103/PhysRevD.73.014019.

33. Ascher, U.; Christiansen, J.; Russell, R. A Collocation Solver for Mixed Order Systems of Boundary Value Problems. *Math. Comput.* **1979**, *33*, 659–679. doi:10.1090/S0025-5718-1979-0521281-7.

34. Kleihaus, B.; Kunz, J.; List, M. Rotating boson stars and Q-balls. *Phys. Rev. D* **2005**, *72*, 064002. doi:10.1103/PhysRevD.72.064002.

35. Brihaye, Y.; Diemer, V.; Hartmann, B. Charged Q-balls and boson stars and dynamics of charged test particles. *Phys. Rev. D* **2014**, *89*, 084048. doi:10.1103/PhysRevD.89.084048.

36. Fukushima, K. Chiral effective model with the Polyakov loop. *Phys. Lett. B* **2004**, *591*, 277–284. doi:10.1016/j.physletb.2004.04.027.

37. Biswal, M.; Digal, S.; Saumia, P. Z_3 meta-stable states in PNJL model. *arXiv* **2019**, arXiv:1907.07981.

38. Nugaev, E.Y.; Shkerin, A. Review of Nontopological Solitons in Theories with $U(1)$-Symmetry. *J. Exp. Theor. Phys.* **2020**, *130*, 301–320. doi:10.1134/S1063776120020077.

Article

Inflation inside Non-Topological Defects and Scalar Black Holes

Yves Brihaye [1], Felipe Console [2] and Betti Hartmann [2,3,4,*]

[1] Département de Mathématique, Université de Mons , Place du Parc, 7000 Mons, Belgium; yves.brihaye@umons.ac.be

[2] Instituto de Física de São Carlos (IFSC), Universidade de São Paulo (USP), CP 369, São Carlos, SP 13560-970, Brazil; felipe.console@usp.br

[3] Institut für Physik, Carl-von-Ossietzky Universität Oldenburg, 26111 Oldenburg, Germany

[4] Department of Physics and Earth Sciences, Jacobs University Bremen, 28759 Bremen, Germany

* Correspondence: bhartmann@ifsc.usp.br

Received: 29 October 2020; Accepted: 25 November 2020; Published: 22 December 2020

Abstract: In this paper, we demonstrate that a phenomenon described as *topological inflation*, during which inflation occurs inside the core of topological defects, has a non–topological counterpart. This appears in a simple set-up containing Einstein gravity coupled minimally to an electromagnetic field as well as a self-interacting, complex valued scalar field. The U(1) symmetry of the model is unbroken and leads to the existence of globally regular solutions, so-called boson stars, that develop a horizon for sufficiently strong gravitational coupling. We also find that the same phenomenon exists for black holes with scalar hair.

Keywords: black holes; scalar fields; boson stars

---

1. Introduction

The theory of General Relativity (GR) is the best tested theory of gravity developed up to date. The classical tests, within the solar system, where the deviations from Newtonian gravity are small, helped to promote GR to one of the fundamental theories of nature. For strong gravitational fields, i.e., when the deviations from Newtonian gravity become important, observations and tests have now become available. An excellent laboratory to consider strong gravitational field effects are black holes (BHs) and neutron stars (NSs). It is a common belief that BHs are "simpler" to describe than NSs partly due to the lack of knowledge of the equation of state of the matter making up the NS. BHs, in fact, are still believed to follow the so-called no-hair conjecture [1], which states that all stationary asymptotically flat BHs are fully characterized by global charges associated with a Gauss law. There are numerous counter-examples for the conjecture when considering nonlinear matter fields. However, the situation for scalar fields is very different. A number of no-scalar-hair theorems were put forward (see [2] for a recent review) and scalar fields, apparently, should be trivial around a stationary and asymptotically flat BH spacetime.

One process where a scalar field can become non-trivial in a BH spacetime is through spontaneous scalarization. It was first discussed in a scalar-tensor theory of gravity around a NS [3], in which the scalar field couples to the trace of the energy-momentum tensor and can obtain non-vanishing values even if its asymptotic value is zero. The corresponding phenomenon has gained considerable attention in black hole physics [4].

The view on no-hair theorems for minimally coupled scalar fields has changed since the discovery of hairy Kerr BHs in a model where a massive complex scalar field is minimally coupled to gravity [5,6]. To circumvent the no-scalar-hair theorems, it is necessary to assume a harmonic dependence on the time and azimuth coordinates. Then, the so-called synchronization condition $\omega/m = \Omega_H$ must be imposed, where ω is the scalar field frequency, m an integer, and Ω_H the horizon angular velocity. The configuration of the test scalar field outside the event horizon was dubbed "scalar cloud" and when taking backreaction into account leads to the existence of hairy Kerr BH solutions. As the horizon radius approaches zero, the solution is reduced to a spinning *boson star*.

Boson stars (BSs) are regular, stationary, and localized solutions to the Einstein–Klein–Gordon system of equations, formed by a complex scalar field with a continuous U(1) symmetry, which gives rise to a globally conserved Noether charge. They are the self-gravitating counterparts of Q-balls [7].

Their size can range from the atomic scale up to the size of supermassive BHs, depending on the choice of the scalar potential (see [8] for a review), and can be used as models for dark matter particles [9] and BH mimickers [10], for example. The absence of an event horizon, however, can lead to significant changes in the propagation of light rays when compared to the spacetime of a BH [11]. Charged BSs were studied in [12] for a self-interacting scalar potential whose motivation comes from supersymmetric extensions of the standard model and, originally, uncharged Q-balls had been discussed [13–15]. A combination of the attractive effects of gravity and the repulsive effects of electromagnetism can render the charged BS stable. Most work that deals with boson stars is concerned with infinitely extended boson stars, e.g., also when considering these compact objects as black hole mimickers. Moreover, the boson stars that are actually compact do need a very specific potential that is not differentiable at vanishing scalar field value; see, e.g., [16]. In this latter case, the exterior of the boson star would simply be given by the Schwarzschild solution with a scalar field identically zero. Thus, in this paper, we consider only the boson stars with scalar field falling off exponentially at infinity.

Further studies concerning scalar clouds were considered subsequently for Reissner–Nordström (RN) spacetime [17], where, for a non-trivial configuration of the gauged scalar field, it was shown that it is necessary to add self-interactions in the scalar potential and the resonance condition to be satisfied, $\omega = qV(r_h)$, where q is the scalar coupling constant and $V(r_h)$ the electric potential on the horizon.

Gauged scalar clouds in the Schwarzschild BH were considered in [17,18]. In this case, the background is fixed by the Schwarzschild metric and the differential equations for the electric potential and the scalar field are coupled. It was shown that the scalar clouds exist for some range in the gauge coupling, and it was also found numerically in [18] that two different solutions exist for the same values of the gauge coupling constant and the electric potential at infinity. When backreaction is taken into account, the solutions exist up to a maximal value of the gravitational constant and lead to two distinct situations: (i) an extremal BH with a diverging derivative of the scalar field at the horizon and (ii) a RN–de Sitter solution with a screened electric charge.

In this paper, we extend the results of [18] to include globally regular space-times with the same matter field content. The corresponding solutions are charged Q-balls (in a Minkowski space-time) and boson stars. In the following, we will demonstrate that the phenomenon described above doesn't depend on the details of the scalar self-interaction or on the fact that the space-time possesses a priori a horizon. Our paper is organized as follows: in Section 2, we discuss the model and equations of motion, while Section 3 contains our results on black hole and globally regular space-times, respectively. We end with a discussion in Section 4.

2. Model

Here, we will discuss solutions to the following (3+1)-dimensional model :

$$S = \int d^4x \sqrt{-g}\,\mathcal{L} \quad , \quad \mathcal{L} = \frac{\mathcal{R}}{16\pi G} - (D_\mu \Psi)^* D^\mu \Psi - \frac{1}{4} F_{\mu\nu} F^{\mu\nu} - U(|\Psi|) . \tag{1}$$

This is the Einstein–Hilbert action with $\mathcal{R}$ the Ricci scalar and G Newton's constant minimally coupled to a complex valued scalar field Ψ that is charged under a U(1) gauge field A_μ. $D_\mu = \partial_\mu - iqA_\mu$ is the covariant derivative of the scalar field and $F_{\mu\nu} = \partial_\mu A_\nu - \partial_\nu A_\mu$ the field strength tensor of the U(1) gauge field. The scalar field potential $U(|\Psi|)$ will turn out to be a crucial ingredient in our construction in the following. We will choose the potential as follows [13,14] :

$$U(|\Psi|) = \mu^2\eta^2 \left[1 - \exp\left(-\frac{|\Psi|^2}{\eta^2}\right)\right] , \tag{2}$$

where μ corresponds to the mass of the scalar field and η is an energy scale.

In the following, we are interested in spherically symmetric and stationary solutions. The Ansatz for the metric is:

$$ds^2 = -N(r)(\sigma(r))^2 dt^2 + \frac{1}{N(r)}dr^2 + r^2(d\theta^2 + \sin^2\theta d\varphi^2) , \quad N(r) = 1 - \frac{2m(r)}{r} , \tag{3}$$

while the matter fields are chosen according to :

$$\Psi(r,t) = \eta e^{i\tilde{\omega}t}\psi(r) , \quad A_0 = \eta V(r) \tag{4}$$

with $\tilde{\omega}$ a real constant. Note that, although the scalar field is time-dependent, the associated energy-momentum tensor is static and hence likewise the space-time. Defining the following dimensionless quantities

$$x = \mu r , \quad \omega = \frac{\tilde{\omega}}{\mu} , \quad e = \frac{\eta q}{\mu} , \quad \alpha = 4\pi G\eta^2 \tag{5}$$

the equations resulting from the variation of the action (1) depend only on e and α and read (with the prime denoting derivative with respect to x) :

$$m' = \alpha x^2 \left[\frac{V'^2}{2\sigma^2} + N\psi'^2 + U(\psi) + \frac{(\omega - eV)^2}{N\sigma^2}\psi^2\right] \tag{6}$$

$$\sigma' = 2\alpha x\sigma\left[\psi'^2 + \frac{(\omega - eV)^2}{N^2\sigma^2}\psi^2\right] \tag{7}$$

for the metric functions and

$$V'' + \left(\frac{2}{x} - \frac{\sigma'}{\sigma}\right)V' + \frac{2(\omega - eV)\psi^2}{N} = 0 \tag{8}$$

$$\psi'' + \left(\frac{2}{x} + \frac{N'}{N} + \frac{\sigma'}{\sigma}\right)\psi' + \frac{(\omega - eV)^2\psi}{N^2\sigma^2} - \frac{1}{2N}\frac{dU}{d\psi} = 0 . \tag{9}$$

for the matter fields. As is obvious, these equations depend only on the combination $\omega - eV(x)$. Note that the Lagrangian density and with it the energy density are now given in units of $\eta^2\mu^2$.

The asymptotic behaviour of the metric and matter field functions is:

$$N(x \gg 1) = 1 - \frac{2M}{x} + \frac{\alpha Q^2}{x^2} + \ldots\ldots , \quad \sigma(x \gg 1) = 1 + \mathcal{O}(x^{-4}) , \tag{10}$$

$$V(x) = V_\infty - \frac{Q}{x} + \ldots. , \quad \psi(x \to \infty) \sim \frac{\exp(-\mu_{\text{eff},\infty}x)}{x} + \ldots. , \quad \mu_{\text{eff},\infty} = \sqrt{1 - \Omega^2} , \tag{11}$$

where $\Omega^2 := (\omega - eV_\infty)^2$. M and Q denote the (dimensionless) mass and electric charge of the solution, respectively. V_∞ can be understood to be a "chemical potential", i.e., the resistance of the system against the addition of extra charges e to the system. Note that the scalar field possesses an effective mass μ_{eff} smaller than the bare mass of the scalar field, which—in our dimensionless units—is equal to unity.

Next to the electric charge, the solutions possess a Noether charge which results from the unbroken U(1) symmetry of the system. This reads :

$$Q_N = \int dx \, \frac{2x^2 ev\psi^2}{N\sigma} \, .$$

(12)

This quantity can be interpreted as the number of scalar bosons that make up the solution. For globally regular solutions $eQ_N \equiv Q$, while, for black holes, $Q = eQ_N - E_x(x_h)x_h^2/\sigma(x_h)$, where $E_x(x) = -V'(x)$ is the radial electric field of the solution. The second term in this equality is the horizon electric charge and is the consequences of the fact that the horizon corresponds to a surface on which boundary conditions have to be imposed. Finally, the temperature T_H and entropy S of the black hole solutions are given by

$$T_H = (4\pi)^{-1}\sigma(x_h)N'|_{x=x_h} \, , \quad S = \pi x_h^2 \, .$$

(13)

The energy-density ϵ, radial pressure p_r, and tangential pressure $p_\theta = p_\varphi$, respectively, read :

$$\epsilon = -T_t^t = \epsilon_1 + \epsilon_2 + \epsilon_3 + U(\psi)$$

(14)

$$p_r = T_r^r = -\epsilon_1 + \epsilon_2 + \epsilon_3 - U(\psi)$$

(15)

$$p_\theta = T_\theta^\theta = \epsilon_1 - \epsilon_2 + \epsilon_3 - U(\psi) \, .$$

(16)

with

$$\epsilon_1 = \frac{V'^2}{2\sigma^2} \, , \quad \epsilon_2 = N\psi'^2 \, , \quad \epsilon_3 = \frac{(\omega - eV)^2\psi^2}{N\sigma^2} \, .$$

(17)

3. Results

In the following, we will discuss the different types of solutions to the equations of motions (7)–(9). We will start with the discussion of black holes, for which we have integrated the equations between $x = x_h$, i.e., the horizon and infinity, respectively. The solutions correspond to Schwarzschild black holes endowed with charged scalar clouds, so-called Q-clouds, when neglecting the backreaction of the cloud on the space-time, otherwise to charged black holes with scalar hair. We will also discuss globally regular solutions that exist on $x \in [0 : \infty)$. Without and with backreaction, these are charged Q-balls and charged boson stars, respectively.

3.1. Black Hole Solutions

Here, we will discuss space-times that contain a horizon. We begin by discussing Q-clouds on Schwarzschild black holes and will demonstrate that the results obtained in [18] are not specific to the form of the scalar potential and that the existence of two branches of Q-cloud solutions can also be observed when considering a potential of the form (2). We will then discuss the backreaction of these two different branches of solutions onto the space-time.

3.1.1. Solutions without Scalar Fields

Without scalar fields $\psi \equiv 0$, the equations of motion have well-known black hole solutions. These are the uncharged Schwarzschild ($Q = 0$) and the charged Reissner–Nordström solution, respectively:

$$N(x) = 1 - \frac{2M}{x} - \frac{\alpha Q^2}{x^2} \, , \quad \sigma \equiv 1 \, , \quad V(x) = \frac{Q}{x_h} - \frac{Q}{x} \, .$$

(18)

The horizon(s) of this space-time are $x_\pm = M \pm \sqrt{M^2 - \alpha Q^2}$. For the Reissner–Nordström solution, there exists a so-called *extremal limit*, the limit of maximal possible charge for the black hole, which is $x_+ = x_- = M = \sqrt{\alpha}Q$. Note that we have chosen $V(x_h) = 0$ here, such that $V_\infty = Q/x_h$.

3.1.2. Q-Clouds on Schwarzschild Black Holes

In this section, we consider the equations of the matter fields in the background of a Schwarzschild black hole, i.e., we set $\alpha = 0$ and let $\sigma \equiv 1$, $N = 1 - x_h/x$, where x_h is the event horizon radius. In order to ensure regularity of the matter fields on the horizon, we need to impose :

$$\psi'(x_h) = \frac{x_h}{2}\frac{dU}{d\psi}(\psi(x_h)) \ , \quad V(x_h) = 0 \ , \quad \psi(x \to \infty) = 0 \ , \quad V(x \to \infty) = V_\infty \ . \tag{19}$$

Note that $V(x_h) = 0$ is a gauge choice implying $\omega = 0$ and hence $\Omega = eV_\infty$. This is the so-called synchronization condition necessary for the scalar field to be non-trivial in the black hole background.

In Figure 1 (left), we give the value of the scalar field ψ as well as the radial electric field E_x on the horizon for $x_h = 0.15$ and two different values of e dependent on eV_∞. We also give the electic charge Q as well as the electric charge contained in Q_N charges e in this figure (right). This demonstrates that charged Q-clouds exist on a finite interval of $\Omega \in [\Omega_{\min} : 1]$ and that, for each choice of eV_∞, two solutions with different mass M, electric charge Q, and Noether charge Q_N exist. On the first branch of solutions, starting at $eV_\infty = 1$ (which is the threshold for the electric potential to be able to create scalar particles of mass unity) and decreasing eV_∞, the scalar field on the horizon $\psi(x_h)$ increases up to a maximal value. With it, the radial electric field on the horizon, $E_x(x_h)$, decreases in absolute value. Moreover, the electric charge Q and the Noether charge Q_N increase. eV_∞ can be decreased to a minimal value that depends on e and increases with the increase of e. Increasing eV_∞ again from this minimal value, a second branch of charged Q-clouds exist that extends all the way back to $eV_\infty = 1$. On this second branch, the scalar field on the horizon decreases and with it the absolute value of the electric field on the horizon when increasing eV_∞. Interesting, we observe that on this second branch of solutions nearly all electric charge seems to be contained in the Q_N scalar bosons which each carry charge e, while the horizon electric charge decreases indicated by the decreasing electric field on the horizon. Hence, moving along the branches stores increasingly electric charge in the scalar cloud and moves it away from the black hole horizon. Figure 1 (right) further demonstrates that eQ_N is approximately equal to Q on the second branch, while on the first branch we find $eQ_N < Q$.

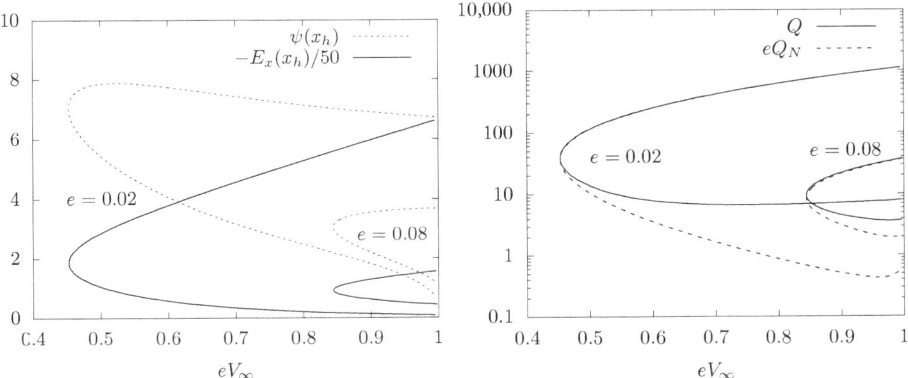

Figure 1. Left: We show the value of the scalar field ψ (dashed) and the radial electric field E_x (solid) on the horizon dependent on $eV_\infty \equiv \Omega$ for $x_h = 0.15$ and two different values of e. **Right**: We show the values of the electric charge Q (solid) and the charge contained in Q_N charges e, eQ_N, (dashed) dependent on $eV_\infty \equiv \Omega$ for the same solutions.

In Figure 2 (left), we show the profiles of the scalar field function $\psi(x)$ and the gauge field function $V(x)$ for the two Q-cloud solutions available for $e = 0.08$, $eV_\infty \equiv \Omega = 0.94$ and $x_h = 0.15$.

When fixing V_∞ and varying e, we find that—again—two branches of solutions exist. These two branches join at a minimal value of the gauge coupling, $e_{\min}$, and exist both up to $e_{\max} = 1/V_\infty$ where the effective mass $\mu_{\mathrm{eff},\infty}$ of the scalar field becomes zero. The minimal value of e has to be determined numerically and we find that $e_{\min} \approx 0.088$ for $V_\infty = 10$ and $e_{\min} \approx 0.109$ for $V_\infty = 8.8$, respectively. In other words, the potential difference between the horizon and infinity, i.e., the chemical potential, needs to be large enough to support the scalar cloud.

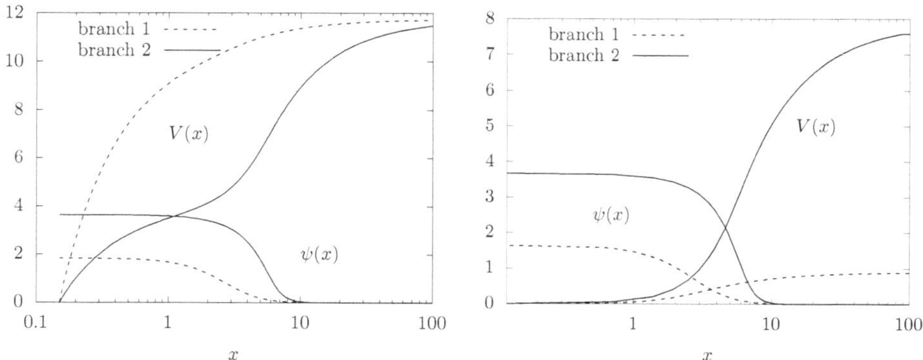

Figure 2. Left: Profiles of the scalar field $\psi(x)$ and the gauge potential $V(x)$ for the two possible Q-cloud solutions on Schwarzschild black holes with $x_h = 0.15$, $e = 0.08$, and $eV_\infty = \Omega = 0.94$. **Right**: Profiles of the scalar field $\psi(x)$ and the gauge potential $V(x)$ for the two possible Q-ball solutions for $e = 0.08$ and $\Omega = 0.94$.

Another interesting question is to understand how the surface area, i.e., the entropy of the background Schwarzschild black hole influences the observations we have made. We have hence fixed e and V_∞ and studied the solutions for varying event horizon radius x_h. Our results are shown in Figure 3, where we give the electric charge Q as well as the electric charge contained in the Q_N scalar bosons that make up the cloud dependent on the event horizon radius x_h. Again, we obtain two branches of solutions. We find that the black hole needs to be sufficiently small in order to allow for the scalar clouds to exist. For the particular choice of parameters here, we find that the maximal possible event horizon radius is $x_h \approx 1.2$. Remembering the rescaling (5), this means that $r_h < 1.2\lambda_{C,\mu}$, where $\lambda_{C,\mu} = 1/\mu$ is the Compton wavelength of the bare scalar field. This implies that we would need an *ultra-light scalar field* in order for the phenomenon of Q-cloud formation to be relevant for astrophysical black holes. e.g., for $\mu = 10^{-10}$ eV, the maximal possible event horizon radius for Q-clouds to exist would be ≈ 2.4 km, the approximate size of a solar mass black hole. We find that, when varying eV_∞, the maximal radius can double or triple, However, it stays the same order of magnitude.

As a final remark, let us mention that, in the limit $x_h \to 0$, we find that $Q \to eQ_N$. This is not surprising since globally regular counterparts to the solutions discussed above exist. These are charged Q-balls (without backreaction of the space-time) and charged boson stars (with backreaction), respectively. Moreover, Figure 3 indicates that, also in the regular limit, two different types of solutions should exist when fixing e and V_∞. This is what we will discuss in Section 3.2.

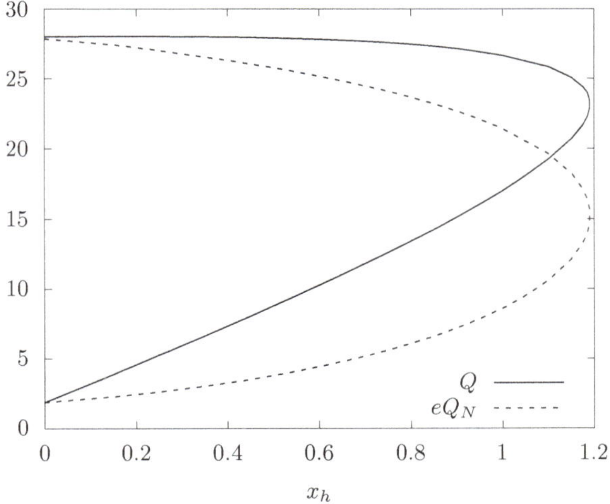

Figure 3. We show the electric charge Q and the electric charge contained in Q_N charges e of Q-clouds on Schwarschild black holes with $x_h = 0.15$ and for $e = 0.08$, $V_\infty = 11.75$, i.e., $\Omega = eV_\infty = 0.94$.

3.1.3. Backreaction of Q-Clouds

For $\alpha > 0$, the Q-clouds backreact on the space-time and modify the original Schwarzschild black hole. The result is a charged black hole which carries scalar hair on its horizon. Note that the existence of these solutions does not contradict no-hair theorems due to the specific form of the scalar field potential. In [18], the same model as in this paper has been discussed, however, for a ψ^6-self-interaction potential. It was noticed that, when approaching the critical value of the gravitational coupling for solutions on the second branch of solutions (those with larger values of $\psi(x)$), the black holes form a second horizon. This second horizon can be interpreted as the horizon of an extremal RN solution. Here, we find that this is also true when considering an exponential-type potential for the scalar field. We show the approach to criticality for the metric and matter field functions in Figure 4 for $\Omega = 0.6$ and $x_h = 0.15$. We find that, at a critical value of $\alpha = \alpha_{cr}$, the solution forms a second horizon at $x = x_h^{(ex)}$ that agrees with the value of an extremal horizon of the corresponding RN solution. Our numerical results indicate that $x_h^{(ex)} \approx M \approx \sqrt{\alpha}Q \approx 121$ for our choice of couplings.

The question then remains how these solutions can be interpreted and why they exist. In fact, this can be understood when considering the energy density components, see (17). In Figure 5 (left), we plot the different components very close to the critical value of $\alpha = \alpha_{cr} \approx 0.000113$. Clearly, the scalar field energy $U(\psi)$ dominates all the other energy momentum components up to a radius $x \approx \tilde{x}$, with $\tilde{x}$ on the order of 100 for this particular choice of α and Ω. This means that the energy-momentum tensor is approximately of the form $T^\nu_\mu = \mathrm{diag}(\epsilon, p, p, p)$, where $\epsilon = -p = U(\psi)$. This is the energy of a perfect fluid that has the equation of state of a positive cosmological constant. In fact, assuming that $U(\psi) = U_0$ dominates the energy density up to $x = \tilde{x}$, we can integrate the equation for m to get

$$N_{cr}(x) \approx 1 - \frac{2}{3}\alpha U_0 x^2 , \qquad (20)$$

while $\sigma = \sigma_0 \neq 0$. This is a space-time with a positive cosmological constant $\Lambda = 2\alpha U_0$. For our choice of potential, $U_0 = 1$. Using $\alpha_{cr} \approx 0.000113$, the cosmological horizon of this solution would be at $x = x_c \approx 115 \approx \tilde{x}$. This is not equal to the value of the extremal horizon $x_h^{(ex)}$. This is different to what is found in the case of inflating monopoles (see e.g., [19,20]) or BH solutions with scalar hair in a scalar-tensor gravity model (see [21] in this issue), where $x_c = x_h^{(\mathrm{ex})}$. We rather find a "transition zone" between $\tilde{x}$ and the extremal RN solution that is of finite thickness—in our dimensionless coordinates, $\Delta x := x_h^{(\mathrm{ex})} - \tilde{x} = \mathcal{O}(10)$.

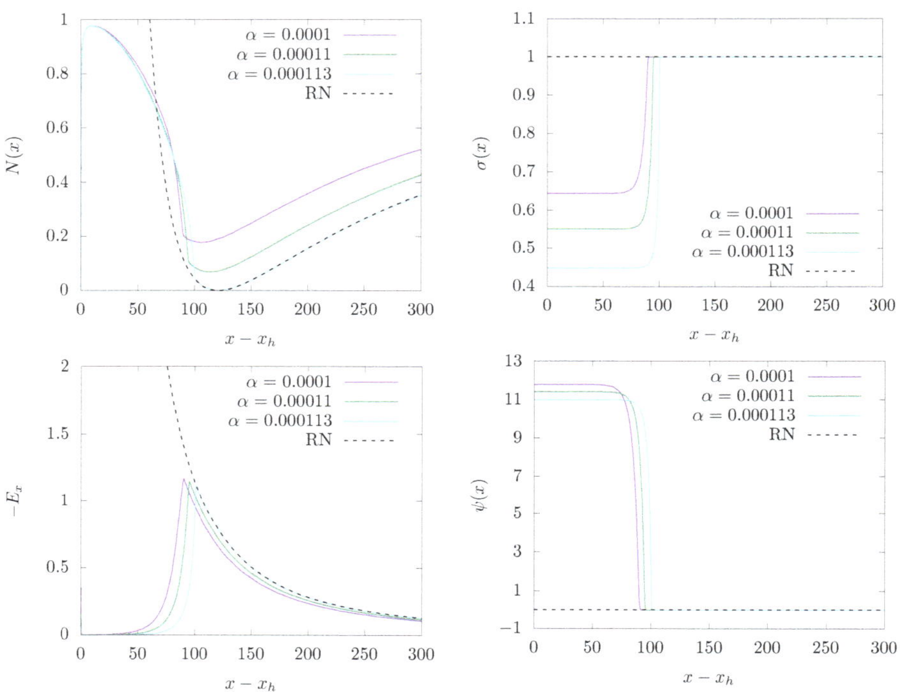

Figure 4. We show the approach to critically for a back hole with charged scalar hair for $\Omega = 0.6$ and $x_h = 0.15$.

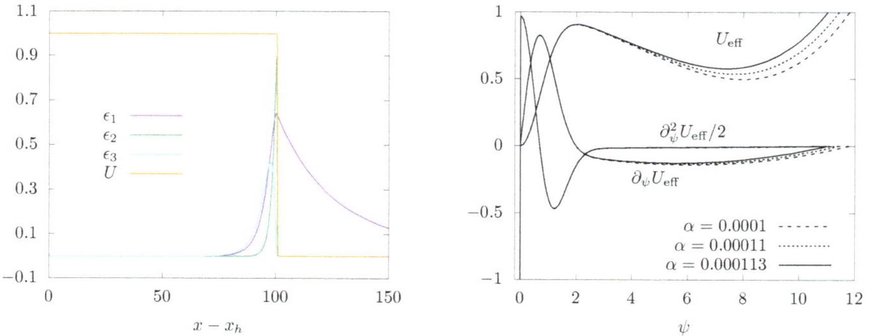

Figure 5. **Left**: We show the profiles of the energy density components (see (17)) for $x_h = 0.15$, $\Omega = 0.6$ and close to the critical value of α, $\alpha_{cr} \approx 0.000113$. **Right**: We show the effective potential $U_{\mathrm{eff}} = U - \frac{(\omega - eV)^2}{N\sigma^2}\psi^2$ and the first ($\partial_\psi U_{\mathrm{eff}}$) and second derivative ($\partial_\psi^2 U_{\mathrm{eff}}$) with respect to ψ at the approach to criticality for $x_h = 0.15$ and $\Omega = 0.6$.

Figure 5 (right) further demonstrates that the reason for the existence of these solutions is related to the coupling of the scalar field to the electromagnetic field. The "bare" potential $U(\psi)$ does not possess extrema away from $\psi = 0$, while the effective potential $U_{\mathrm{eff}} = U(\psi) - \frac{(\omega - eV)^2}{N\sigma^2}\psi^2$ does. For α close to α_{cr} and $\Omega = 0.6$, there exists a local minimum for $\psi = \mathcal{O}(10)$. This is the value that ψ takes on the horizon and up to $x \approx x_c$. Moreover, the first and second derivative of the effective potential with respect to ψ, $\partial_\psi U_{\mathrm{eff}}$ and $\partial_\psi^2 U_{\mathrm{eff}}$, respectively, are smaller than the potential itself close to this value of ψ. Hence, the effective potential is sufficiently "flat" to allow for inflation.

3.2. Globally Regular Solutions

For globally regular solutions, we have to impose boundary conditions at the origin $x = 0$ which ensure regularity. These are

$$m(0) = 0 \ , \ V(0) = 0 \ , \ V'|_{x=0} = 0 \ , \ \psi'|_{x=0} = 0 \ . \tag{21}$$

The condition $V(0) = 0$ is, in fact, a gauge choice. At spatial infinity, the boundary conditions are chosen such that the solution is asymptotically flat and has finite energy. They read:

$$\psi(x \to \infty) \to 0 \ , \ N(x \to \infty) \to 1 \ . \tag{22}$$

Note, in particular, that now $\omega \neq 0$ is necessary for globally regular solutions to exist and that it cannot be set to zero as in the black hole case. In the following, however, we will only use the gauge-invariant quantity $\Omega = \omega - eV_\infty$ to describe the solutions.

3.2.1. (Un)Charged Q-Balls

For $\alpha = 0$, the matter field Equations (8) and (9) decouple from the gravity equations and the metric functions are $m \equiv 0$ (i.e., $N \equiv 1$) and $\sigma \equiv 1$ (or equal to any non-vanishing, positive constant). The remaining Equations (8) and (9) possess non-trivial solutions, so-called *(charged) Q-balls*. For $V \equiv 0$, the solutions for the specific potential (2) have been studied in [13–15], while, for $V \neq 0$, they were investigated in [12]. Let us remind the reader of the most important features of these solutions in the following and add details that are important for the discussion in the following. For $e = 0$, the solutions exist for $\Omega \in (0 : 1]$, and there is a one-to-one relation between the value of the scalar field ψ at the origin, $\psi(0)$, and Ω. The limit $\psi(0) \to 0$ corresponds to $\Omega \to 1$. In this limit, the solution becomes $\psi(x) \equiv 0$, However, neither the mass nor the Noether charge Q_N of the solutions vanish. Increasing the value $\psi(0)$ from zero, Ω decreases, while the mass M and Noether charge Q_N increase monotonically. The dependence of Q_N on Ω for $e = 0$ is given in Figure 6 (left). This demonstrates that, for $\Omega \to 0$, the Noether charge Q_N diverges with $\psi(0)$ tending to infinity. That means that the central density of the uncharged Q-ball is not restricted and can become arbitrarily large.

When choosing $e \neq 0$, charged Q-balls can be constructed. However, there is a very crucial difference to the uncharged case: the central value of $\psi(0)$ is limited to a finite value. As such, solutions exist only for $\Omega \in [\Omega_{\min} : 1]$ with $\Omega_{\min} > 0$ for $e \neq 0$. Moreover, two charged Q-ball solutions—and not one as in the uncharged case—are possible. Figure 6 demonstrates the existence of this second branch of solutions for several values of e. Here, we give the Noether charge Q_N of the (un)charged Q-balls (including the uncharged case $e = 0$) (left) as well as the central value of the scalar field, $\psi(0)$, for several values of e and dependent on Ω. Note that this is qualitatively similar to the case for Q-clouds on Schwarzschild black holes, see Figure 1 (left): decreasing Ω on the first branch of solutions, the value of $\psi(0)$ increases up to a maximal value at $\Omega_{\min} > 0$. From there, a second branch of solutions extends all the way back to $\Omega = 1$. On this second branch, the value of $\psi(0)$ slightly decreases when increasing Ω. Moreover, we find hat $\Omega_{\min}$ is more or less equal when considering charged Q-clouds and Q-balls, respectively. The presence of the horizon in the former case does not seem to influence this value. The Noether charge Q_N of the Q-balls on the second branch is much larger than that of solutions on the first branch. Hence, the Q-balls on the second branch are composed of many more scalar bosons than those on the first branch.

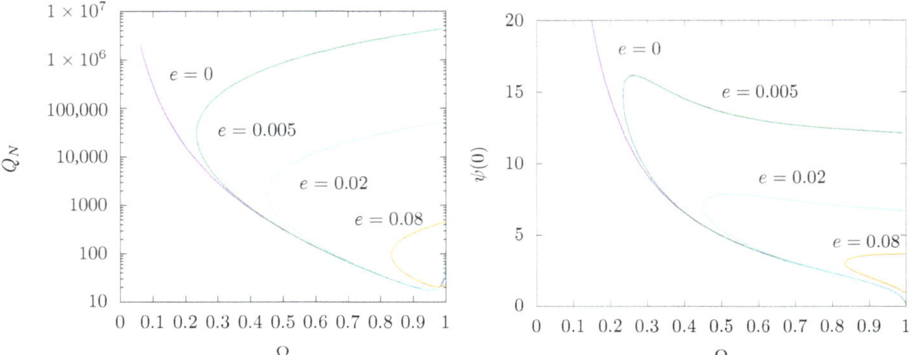

Figure 6. Left: We show the Noether charge Q_N of (un)charged Q-balls dependent on Ω for four different values of e. **Right**: We show the central value of the scalar field, $\psi(0)$, of (un)charged Q-balls dependent on Ω for four different values of e.

When two solutions exist for the same value of Ω, we will use the convention in the following to denote "branch 1" (resp. "branch 2") the branch of solutions with the lower (resp. higher) Noether charge Q_N. The profiles of the scalar field $\psi(x)$ and the electric potential $V(x)$ of the two different Q-ball solutions for $e = 0.08$ and $\Omega = 0.94$ are shown in Figure 2 (right). Although some features appear to be similar when comparing Q-clouds and Q-balls, there is one crucial difference which is apparent from Figure 2: the electric field. For Q-balls, it vanishes at the origin (as it should), while it is non-zero on the horizon of the black hole that carries the charged Q-cloud.

3.2.2. Charged Boson Stars

For $\alpha \neq 0$, charged boson star solutions exist. We have constructed these solutions and found that a previously unnoticed phenomenon is present when the U(1) in the model is gauged, i.e., when the scalar field is charged under the U(1). In fact, this phenomenon is very similar to that observed for charged black holes with scalar hair (see [18] and discussion above). Choosing a solution on the second branch of solutions, where $\psi(x)$ is large at and close to the origin and increasing the gravitational coupling leads to the appearance of a horizon at a critical value of α. The approach to criticality is shown in Figure 7 for $\Omega = 0.6$. When increasing α, we find that the minimal value of $N(x)$ tends to zero, while $\sigma(0)$ decreases (but does not reach zero). For $\alpha = \alpha_{cr}$, we find that, outside a given value of the radial coordinate $x = \tilde{x}$, the solution corresponds to the extremally charged Reissner–Nordström solution with $\sigma \equiv 1$ and $\psi \equiv 0$. For $\Omega = 0.6$, we find that $\alpha_{cr} \approx 0.000113$. Note that the value of $\tilde{x}$ and the location of the extremal horizon at $x_h^{(ex)} \approx 121$ do not coincide in this case. In agreement with our interpretation of the limiting solution, we find that the mass at α_{cr} is $M = x_h^{(ex)} = \sqrt{\alpha}Q \approx 121$. Comparing this to the case of charged black holes with scalar hair, we find that fixing Ω leads to the same numerical values of α_{cr}, x_c and $x_h^{(ex)}$ for both cases. This suggests that the phenomenon of inflation observed for black holes has a regular limit for $x_h \to 0$. The reason for this is also connected to the observation that, while for boson stars we always have $eQ_N = Q$, this is not true for black hole. However, for the black hole solutions on the second branch, we find $eQ_N \approx Q$, which means that (essentially) all electric charge is carried by the scalar field. Hence, these black holes behave very similar to charged boson stars.

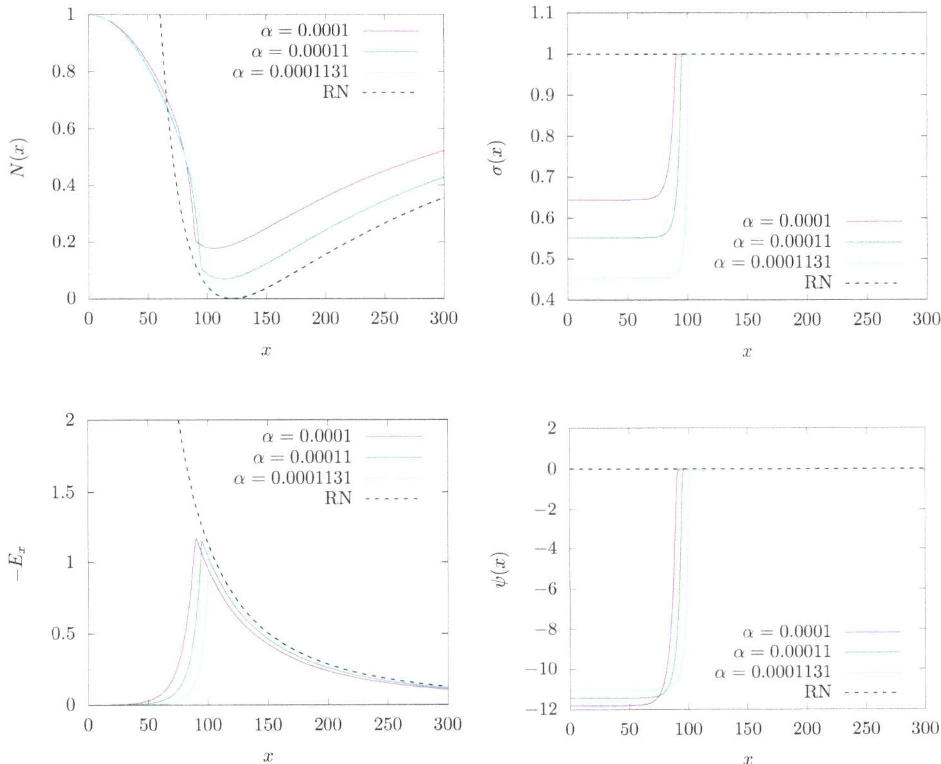

Figure 7. We show the approach for a charged boson star with $\Omega = 0.6$.

The interpretation of the solutions is, hence, very similar, although we are now dealing with an a priori *globally regular* space-time. In particular, for the same choice of Ω and α, Figure 5 (left and right) is identical to the corresponding equivalent plots for the boson stars. Note that, while the component ϵ_1 is zero for the boson stars *by construction*, it is zero for the black hole in this particular limit. To re-emphasize this point, only in this particular limit does the electric field on the horizon of the black hole vanish indicating that the horizon electric charge is vanishing.

As we have discussed above, the exterior of the black hole is hence inflating with the scalar field energy dominating the remaining energy-momentum components and hence driving inflation. We find the interior of the boson star defined by $U(\psi) = U_0 \approx 1$ (as compared to the exterior of the boson star, where $U(\psi) = 0$) is inflating. Hence, *these solutions are the non-topological equivalent of the inflating topological defects (such as magnetic monopoles) that have been discussed in [22,23].* These latter defects contain a core that is trapped inside the false vacuum of the model and this is exactly what provides the vacuum energy to drive inflation. Consequently, in these models of *topological inflation*, topological defects can "blow up" to sizes of the universe. Here, we find that a similar phenomenon happens for localized, globally regular solutions made up out of a complex valued scalar field coupled to an Abelian gauge field.

4. Discussion

In this paper, we have studied black holes and globally regular solutions in a simple scalar field model with scalar field self-interaction of an exponential type that is motivated from models of gauge-mediated supersymmetry breaking. The crucial point for the existence of the solutions presented here is the self-interacting of the scalar field as well as the coupling to the electromagnetic field. This latter coupling allows for the effective scalar field potential to possess a local minimum (the "false vacuum") in which the scalar field can become trapped for specific choices of the coupling constants such that the energy density of these solutions becomes dominated by the scalar field energy of this false vacuum. When letting these solutions backreact on the space-time, we find that, for sufficiently strong gravitational coupling, a second horizon starts to form which corresponds to the horizon of an extremal RN solution.

In order to get an idea of whether these results could be important at cosmological scale, let us go back to dimensional couplings for a moment. For $e = 0.005$ (corresponding to the case $\Omega = 0.6$ described above), we find that $\alpha_{cr} \approx 0.000113$. This corresponds to $\eta \approx 0.003 M_{Pl}$, where $M_{Pl} = G^{-1/2}$. Hence, the energy scale of the scalar field would have to be on the order of the Grand Unification scale. The remaining details depend very much on the mass μ of the scalar field. $e = 0.005$ then implies $q \approx 1.67(\mu/M_{Pl})$ and a Hubble constant associated with the expansion of $H \sim \sqrt{\frac{8\pi G U_0}{3}} = \sqrt{\frac{8\pi}{3} \frac{\mu\eta}{M_{Pl}}} \approx 0.0087\mu$.

Let us now discuss the new phenomenon observed here compared to the topological inflation scenario. What we have noticed in the model at hand is a phenomenon not discussed in the literature so far. It contains an unbroken U(1) symmetry (in contrast to the model of topological inflation, where a spontaneous symmetry breaking occurs) and is hence much closer to the scalar field models normally used in inflation. Considering topological inflation to occur inside the core of magnetic monopoles typically requires a higher rank gauge group to be broken to U(1) and hence a Higgs field in the adjoint representation of the group. All of this is not necessary in our case. We are dealing with a simple set up that does not require any symmetry breaking mechanism, However, only the coupling to a U(1) gauge field. Certainly, what we have presented is a toy model, However, we believe that it is quite generic.

5. Materials and Methods

We have used the Black box solver COLSYS.

Author Contributions: Conceptualization, Y.B. and B.H.; methodology, Y.B. and B.H.; software, Y.B.; validation, Y.B. and B.H.; formal analysis, Y.B. and B.H.; investigation, Y.B., F.C., and B.H.; resources, Y.B. and B.H.; data curation, Y.B. and B.H.; writing—original draft preparation, B.H. and F.C.; writing—review and editing, Y.B., F.C., and B.H.; visualization, B.H.; supervision, Y.B. and B.H.; project administration, Y.B. and B.H.; funding acquisition, B.H. All authors have read and agreed to the published version of the manuscript.

Funding: This research was funded by FAPESP under Grant No. 2019/01511-5 and the DFG under the RTG 1620 *Models of gravity*. This research was also financed in part by the Coordenação de Aperfeiçoamento de Pessoal de Nível Superior - Brasil (CAPES) under Finance Code 001.

Acknowledgments: B.H. would like to thank FAPESP for financial support under grant 2019/01511-5 as well as the DFG Research Training Group 1620 Models of Gravity for financial support.

Conflicts of Interest: The authors declare no conflict of interest.

Abbreviations

The following abbreviations are used in this manuscript:

BH	Black Hole
BS	Boson Star
NS	Neutron Star
RN	Reissner–Nordström

References

1. Ruffini, R.; Wheeler, J.A. Introducing the black hole. *Phys. Today* **1971**, *24*, 30. [CrossRef]
2. Herdeiro, C.A.R.; Radu, E. Asymptotically flat black holes with scalar hair: A review. *Int. J. Mod. Phys. D* **2015**, *24*, 1542014. [CrossRef]

3. Damour, T.; Esposito-Farese, G. Nonperturbative strong field effects in tensor-scalar theories of gravitation. *Phys. Rev. Lett.* **1993**, *70*, 2220. [CrossRef] [PubMed]
4. Silva, H.O.; Sakstein, J.; Gualtieri, L.; Sotiriou, T.P.; Berti, E. Spontaneous scalarization of black holes and compact stars from a Gauss-Bonnet coupling. *Phys. Rev. Lett.* **2018**, *120*, 131104. [CrossRef]
5. Hod, S. Stationary Scalar Clouds Around Rotating Black Holes. *Phys. Rev. D* **2012**, 86, 104026. [CrossRef]
6. Herdeiro, C.A.R.; Radu, E. Kerr black holes with scalar hair. *Phys. Rev. Lett.* **2014**, *112*, 221101. [CrossRef]
7. Coleman, S.R. Q-Balls. *Nucl. Phys. B* **1985**, *262*, 263. [CrossRef]
8. Schunck, F.E.; Mielke, E.W. General relativistic boson stars. *Class. Quant. Grav.* **2003**, *20*, R301. [CrossRef]
9. Eby, J.; Kouvaris, C.; Nielsen, N.G.; Wijewardhana, L.C.R. Boson Stars from Self-Interacting Dark Matter. *JHEP* **2016**, *02*, 028. [CrossRef]
10. Guzman, F.S.; Rueda-Becerril, J.M. Spherical boson stars as black hole mimickers. *Phys. Rev. D* **2009**, *80*, 084023. [CrossRef]
11. Macedo, C.F.B.; Pani, P.; Cardoso, V.; Crispino, L.C.B. Astrophysical signatures of boson stars: Quasinormal modes and inspiral resonances. *Phys. Rev. D* **2013**, *88*, 064046. [CrossRef]
12. Brihaye, Y.; Diemer, V.; Hartmann, B. Charged Q-balls and boson stars and dynamics of charged test particles. *Phys. Rev. D* **2014**, *89*, 084048. [CrossRef]
13. Campanelli, L.; Ruggieri, M. Supersymmetric Q-balls: A Numerical study. *Phys. Rev. D* **2008**, *77*, 043504. [CrossRef]
14. Copeland, E.J.; Tsumagari, M.I. Q-balls in flat potentials. *Phys. Rev. D* **2009**, *80*, 025016. [CrossRef]
15. Hartmann, B.; Riedel, J. Supersymmetric Q-balls and boson stars in (d+1) dimensions. *Phys. Rev. D* **2013**, *87*, 044003. [CrossRef]
16. Hartmann, B.; Kleihaus, B.; Kunz, J.; Schaffer, I. Compact Boson Stars. *Phys. Lett. B* **2012**, *714*, 120. [CrossRef]
17. Herdeiro, C.A.R.; Radu, E. Spherical electro-vacuum black holes with resonant, scalar Q-hair. *Eur. Phys. J. C* **2020**, *80*, 390. [CrossRef]
18. Brihaye, Y.; Hartmann, B. Strong gravity effects of charged Q-clouds and inflating black holes. *arXiv* **2020**, arXiv:2009.08293.
19. Breitenlohner, P.; Forgacs, P.; Maison, D. Gravitating monopole solutions. *Nucl. Phys. B* **1992**, *383*, 357. [CrossRef]
20. PBreitenlohner; Forgacs, P.; Maison, D. Gravitating monopole solutions 2. *Nucl. Phys. B* **1995**, *442*, 126. [CrossRef]
21. Blázquez-Salcedo, J.L.; Kahlen, S.; Kunz, J. Critical solutions in scalarized black holes. *arXiv* **2020**, arXiv:2011.01326.
22. Vilenkin, A. Topological inflation. *Phys. Rev. Lett.* **1994**, *72*, 3137. [CrossRef] [PubMed]
23. Linde, A.D. Monopoles as big as a universe. *Phys. Lett. B* **1994**, *327*, 208. [CrossRef]

Article

Critical Solutions of Scalarized Black Holes

Jose Luis Blázquez-Salcedo [1,2], Sarah Kahlen [2,*] and Jutta Kunz [2]

[1] Departamento de Física Teórica, Facultad de Ciencias Físicas, Universidad Complutense de Madrid, 28040 Madrid, Spain; joseluis.blazquez@fis.ucm.es

[2] Institut für Physik, Universität Oldenburg, Postfach 2503, D-26111 Oldenburg, Germany; jutta.kunz@uni-oldenburg.de

[*] Correspondence: sarah.kahlen@uni-oldenburg.de

Received: 12 November 2020; Accepted: 9 December 2020; Published: 11 December 2020

Abstract: We consider charged black holes with scalar hair obtained in a class of Einstein–Maxwell–scalar models, where the scalar field is coupled to the Maxwell invariant with a quartic coupling function. Besides the Reissner–Nordström black holes, these models allow for black holes with scalar hair. Scrutinizing the domain of existence of these hairy black holes, we observe a critical behavior. A limiting configuration is encountered at a critical value of the charge, where space time splits into two parts: an inner space time with a finite scalar field and an outer extremal Reissner–Nordström space time. Such a pattern was first observed in the context of gravitating non-Abelian magnetic monopoles and their hairy black holes.

Keywords: black holes; scalar fields; Einstein–Maxwell–scalar theory

---

1. Introduction

In recent years, studies of black holes with scalar hair received much interest, both in the context of generalized gravity theories such as Einstein–scalar–Gauß–Bonnet (EsGB) theories [1–22] and in the context of simpler models such as Einstein–Maxwell–scalar (EMs) models [23–40]. In both cases, the way to couple scalar fields to the respective invariants, the Gauß–Bonnet and the Maxwell invariant, is crucial for the resulting types of black hole solutions and their properties. In fact, according to the choice of coupling function, distinct classes arise. In the first class, black holes with scalar hair are present but with no general relativity (GR) black holes. Examples here are dilatonic black holes, first obtained long ago [41,42]. In contrast, the second class allows for both black holes with scalar hair and GR black holes. In the latter case, we should distinguish whether the GR black holes can exhibit tachyonic instability, causing spontaneous scalarization of black holes [1–3,23], or whether the GR black holes will never succumb to such an instability [31].

An example of the latter has been studied in [38,39]. In this set of models, the coupling function $f(\Phi) = 1 + \alpha \Phi^4$ has been chosen, coupling the scalar field ϕ to the Maxwell invariant with coupling constant α. Clearly, this coupling function allows for GR black hole solutions, since the resulting coupled set of field equations can always be satisfied with a vanishing scalar field. Thus, Reissner–Nordström (RN) black holes are solutions of this set of models with their usual domain of existence, limited by the set of extremal RN black holes. However, this set of models allows also for black holes with scalar hair. These scalarized solutions can be found in two different sets or branches: the cold branch and the hot branch [38,39], that are labeled according to their horizon temperature. For fixed coupling α, their domain of existence can be expressed in terms of their charge to mass ratio $q = Q/M$. The cold branch resides in the interval $[q_{\min}(\alpha), 1]$ and the hot branch resides in $[q_{\min}(\alpha), q_{\max}(\alpha)]$, while the bald RN branch resides in $[0, 1]$.

On the other hand, hairy black holes with non-Abelian gauge fields have been studied for a long time (see, e.g., [43–45] for reviews). The Einstein–Yang–Mills (EYM) and Einstein–Yang–Mills–Higgs

(EYMH) theories typically allow not only for black holes with non-Abelian hair but also for embedded Abelian solutions, such as RN black holes. This is thus analogous to the EMs models of the second class. Unlike most EMs models though, the EYM and EYMH models also feature globally regular solutions, solitons. In the SO(3)-EYMH case, these solitons correspond to gravitating magnetic monopoles and dyons, which can be endowed with a horizon, generating hairy black holes [46–52]. These non-Abelian solutions do not exist for arbitrary values of the coupling constant or horizon radius but possess a limited domain of existence. As one of the boundaries is approached, an interesting limiting behavior is observed: At a critical radial coordinate $r_{cr} = Q_{cr}$, the space time divides into two parts. In the exterior part $r > r_{cr}$, the scalar field assumes its vacuum expectation value while the non-Abelian gauge field vanishes except for an embedded Abelian field, yielding the exterior region of an extremal RN black hole with degenerate horizon r_{cr}. In the interior part $r < r_{cr}$, nontrivial non-Abelian and scalar fields remain. These tend to their respective vacuum values at r_{cr}. As this critical solution is approached, both parts of the space time become infinite in extent, since the metric coefficient of the radial coordinate features the double zero of a degenerate horizon.

Here, we will consider the limiting behavior of the EMs hairy black holes on the cold branch as the upper boundary of their domain of existence, $q = 1$, is approached. Since the upper limit indeed has $q = 1$ and a vanishing horizon temperature, one may expect that an extremal RN solution is approached. However, as we will show, this is only part of the truth. In fact, an analogous critical behavior is encountered, as has been known in the case of non-Abelian solutions for long. As the critical solution is approached, the space time splits into two parts, an exterior part $r > r_{cr}$ corresponding to the exterior region of an extremal RN black hole and an interior part $r < r_{cr}$ with a finite scalar field that vanishes at r_{cr}.

In Section 2, we present the EMs theory and the equations of motion. We then recall the Ansatz for spherically symmetric black hole solutions and the expansions at the horizon and at infinity. The properties of the black hole solutions are recalled in Section 3. Next, we consider the critical behavior for fixed coupling constant α and then address the α-dependence of the properties of the critical solutions in the same section. In Section 4, we show that the excited solutions exhibit an analogous critical behavior. We conclude in Section 5.

2. EMs Theory

We consider EMs theory described by the action

$$\mathcal{S} = \int d^4x \sqrt{-g} \left[R - 2\partial_\mu \Phi \partial^\mu \Phi - f(\Phi) F_{\mu\nu} F^{\mu\nu} \right], \tag{1}$$

with the Ricci scalar R, the real scalar field Φ, the Maxwell field strength tensor $F_{\mu\nu}$ and the coupling function $f(\Phi)$. We use units so that $c = G = 1$.

In this paper, we will assume a quartic dependence:

$$f(\Phi) = 1 + \alpha \Phi^4. \tag{2}$$

We will focus on positive values of the coupling constant α, meaning $\Phi = 0$ is the global minimum of the coupling function. This type of coupling is representative of the type IIB (scalarised-disconnected-type following the nomenclature introduced in [31]), since it allows for the existence of both bald (RN) and hairy (scalarised) black holes. However, RN solutions are not unstable with respect to scalar perturbations [38,39].

The Einstein–, Maxwell– and scalar-field equations follow from the variational principle and read as follows:

$$R_{\mu\nu} - \frac{1}{2}g_{\mu\nu}R = T^{\Phi}_{\mu\nu} + T^{EM}_{\mu\nu} \,, \tag{3}$$

$$\nabla_{\mu}\left(\sqrt{-g}f(\Phi)F^{\mu\nu}\right) = 0 \,, \tag{4}$$

$$\frac{1}{\sqrt{-g}}\partial_{\mu}\left(\sqrt{-g}g^{\mu\nu}\partial_{\nu}\Phi\right) = \dot{f}(\Phi)F_{\mu\nu}F^{\mu\nu} \,, \tag{5}$$

with $\dot{f}(\Phi) = df(\Phi)/d\Phi$, electromagnetic stress-energy tensor

$$T^{EM}_{\mu\nu} \equiv 2f(\Phi)\left(F_{\mu\alpha}F_{\nu}{}^{\alpha} - \frac{1}{4}g_{\mu\nu}F^2\right) \tag{6}$$

and scalar stress-energy tensor

$$T^{\Phi}_{\mu\nu} \equiv \frac{1}{2}\partial_{\mu}\Phi\partial_{\nu}\Phi - \frac{1}{2}g_{\mu\nu}\frac{1}{2}(\partial_{\alpha}\Phi)^2 \,. \tag{7}$$

To study static spherically symmetric black holes, we employ the following parametrization of the metric:

$$ds^2 = -N(r)e^{-2\delta}dt^2 + \frac{dr^2}{N(r)} + r^2(d\theta^2 + \sin^2\theta d\varphi^2) \,, \tag{8}$$

with the metric functions $N(r)$ and $\delta(r)$. However, for some expressions, it is useful to define the mass function $m(r)$ as $N(r) = 1 - \frac{2m(r)}{r}$ and the metric function as $g(r) = N(r)e^{-2\delta(r)} = -g_{tt}(r)$.

To obtain black holes with electric charge and, in the scalarized case, also scalar charge, we parametrize the gauge potential and the scalar field by

$$A_{\mu} = \left(A_t(r),0,0,0\right) \,, \quad \Phi = \Phi(r) \,. \tag{9}$$

The above Ansatz for a static and spherically symmetric configuration is substituted into the equations of motion (3)–(5). This leads to the following set of coupled ordinary differential equations (ODEs) for the unknown functions of the metric and the fields:

$$m' = \frac{r^2N\Phi'^2}{2} + \frac{Q^2}{2r^2f(\Phi)} \,, \delta' + r\Phi'^2 = 0 \,, A'_t = -\frac{Qe^{-\delta}}{f(\Phi)r^2} \,, \tag{10}$$

$$\Phi''(r) + \frac{1+N}{rN}\Phi' - \frac{Q^2}{r^3Nf(\Phi)}\left(\Phi' - \frac{\dot{f}(\Phi)}{2rf(\Phi)}\right) = 0 \,, \tag{11}$$

where a prime denotes a derivative with respect to the radial coordinate and Q is the electric charge of the black holes.

To address the vicinity of the black hole horizon, we perform a power series expansion of the functions of the metric and the fields in $r - r_H$ at the horizon, where we denote the horizon radius by r_H and the horizon values of the functions by the subscript H:

$$m(r) = \frac{r_H}{2} + m_1(r - r_H) + \cdots \,, \qquad \delta(r) = \delta_H - \Phi_1^2 r_H(r - r_H) + \cdots \,, \tag{12}$$

$$A_t(r) = \Psi_H - \frac{e^{-\delta_H}Q}{r_H^2 f(\Phi_H)}(r - r_H) + \cdots \,, \qquad \Phi(r) = \Phi_H + \Phi_1(r - r_H) + \cdots \,, \tag{13}$$

with

$$m_1 = \frac{Q^2}{2r_H^2 f(\Phi_H)} \,, \quad \Phi_1 = \frac{Q^2\dot{f}(\Phi_H)}{2r_Hf(\Phi_H)\left[Q^2 - r_H^2 f(\Phi_H)\right]} \,. \tag{14}$$

Some important properties of the horizon that we will use are the horizon area $A_H = 4\pi r_H^2$ and the horizon temperature $T_H = N'(r_H)e^{-\delta(r_H)}/4\pi$ (which is determined by the surface gravity at the horizon in the usual way).

A power series expansion in $1/r$ at infinity of the functions of the metric and the fields allows us to read off the global charges of the black holes:

$$m(r) = M - \frac{Q^2 + Q_s^2}{2r} + \cdots , \qquad \delta(r) = \frac{Q_s^2}{2r^2} + \cdots , \tag{15}$$

$$A_t(r) = -\frac{Q}{r} + \cdots , \qquad \Phi(r) = \frac{Q_s}{r} + \frac{MQ_s}{r^2} + \cdots , \tag{16}$$

with Arnowitt-Deser-Misner mass M and scalar charge Q_s, both of them also defined in the standard way.

3. Limit of Cold Black Holes

3.1. Branches of Black Holes

We now briefly recall the properties of static spherically symmetric electrically charged black hole solutions with quartic coupling function (2). The black holes of the RN branch are given by

$$\delta(r) = 0 , \; m(r) = M - \frac{Q^2}{2r} , \; A_t(r) = -\frac{Q}{r} , \; \Phi(r) = 0 . \tag{17}$$

The scalarized black hole solutions are obtained numerically [38]. We solve the field equations subject to the boundary conditions that follow from the above expansions at the horizon and at infinity, with input parameters α, r_H and Q. We employ the professional solver COLSYS [53], which is based on a collocation method for boundary-value ODEs and on a damped Newton method of quasi-linearization. Since this solver includes an adaptive mesh selection procedure, it is very suitable for the problem at hand, where high accuracy is needed in a very small interval close to the black hole horizon. Consequently, the grid is successively refined until the required accuracy is reached, typically 10^{-16}.

The solutions are characterized by a set of dimensionless quantities: the charge to mass ratio q, the reduced horizon area a_H and the reduced horizon temperature t_H, for which

$$q \equiv \frac{Q}{M} , \qquad a_H \equiv \frac{A_H}{16\pi M^2} = \frac{r_H^2}{4M^2} , \qquad t_H \equiv 8\pi M T_H = 2MN'(r_H)e^{-\delta(r_H)} . \tag{18}$$

We illustrate the domain of existence of the black holes in Figure 1a, where we show the mass to charge ratio q versus the coupling constant α. The horizontal black line $q = 1$ represents the set of extremal RN black holes and, thus, the upper boundary for RN black holes. At the same time, that line represents the set of critical scalarized solutions forming the upper boundary of the cold black holes, which reside in the lower green area. The solid green line marks the bifurcation line $q_{\min}(\alpha)$ separating the cold and hot black holes. The hot black holes then extend from this bifurcation line to the dashed red critical line $q_{\max}(\alpha)$, i.e., they fill the whole shaded region.

In Figure 1b, we exhibit the reduced area a_H and reduced temperature t_H (inset) versus the charge to mass ratio q of the black hole solutions for the particular coupling $\alpha = 200$ [38]. The RN branch is shown in solid black, the cold branch is shown in dotted blue and the hot branch is shown in dashed red. Along the cold branch, the mass to charge ratio q decreases, while the reduced area a_H and temperature t_H increase. At the minimal value $q_{\min}$, the cold branch bifurcates with the hot branch. Along the hot branch, a_H decreases again, while t_H increases with increasing q.

From the figure, it seems that the cold branch starts from an extremal RN black hole. Clearly, the charge to mass ratio at its endpoint agrees with the ratio $q = 1$ of an extremal RN black hole and the horizon temperature vanishes at its endpoint, $T_H = 0$. Looking at the horizon area (Figure 1b) and further properties, one is indeed tempted to conclude that the cold branch emerges from an extremal

RN black hole. However, as we will demonstrate in the following, this is only partially true. In fact, the endpoint of the cold branch is a more intriguing configuration.

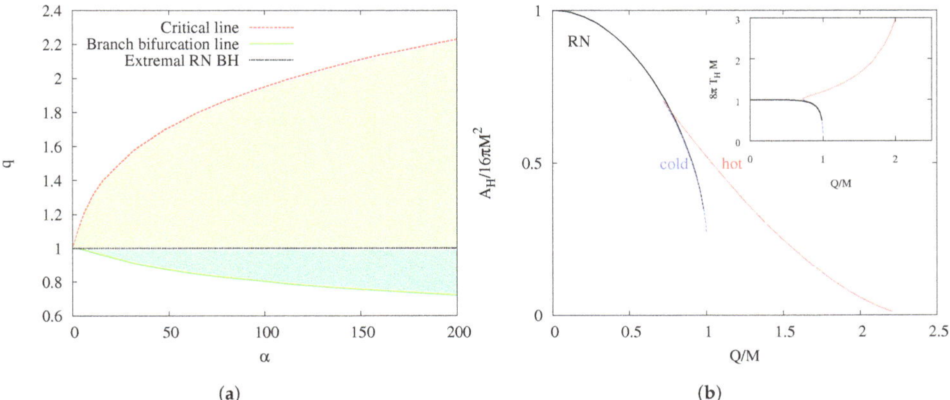

(a)   (b)

Figure 1. (a) Phase diagram of scalarized black holes: mass to charge ratio q vs. coupling constant α. (b) Reduced horizon area a_H (inset: reduced temperature t_H) vs. q for $\alpha = 200$: cold branch (dotted blue), hot branch (dashed red) and RN branch (solid black).

3.2. Approach to the Critical Solution for $\alpha = 200$

To gain an understanding of the limiting configurations, we now inspect a family of cold black holes with fixed coupling constant $\alpha = 200$, fixed horizon radius $r_H = 2$ and increasing charge Q in order to reach the critical solution with $q = 1$ at some Q_{cr}. If the family of cold black holes simply approached an extremal RN black hole, this extremal black hole would have its degenerate horizon at $r_H = 2$, and it would satisfy $r_H = Q = M$.

In Figure 2, we exhibit a set of these cold black hole solutions as they approach the endpoint of the cold branch. Notably, all the solutions exhibited here possess a charge $Q \geq 2$, i.e., they range between the extremal RN limit of $Q = 2$ and a critical value $Q_{cr} = 2.0016701 \, (+O(10^{-8}))$. Thus, they slightly exceed the extremal RN limit for fixed horizon radius $r_H = 2$. The figure shows the metric functions $g(r)$ (a) and $N(r)$ (b), the electromagnetic function $A_t(r)$ (c) and the scalar field function $\Phi(r)$ (d) versus the compactified radial coordinate $x = 1 - \frac{r_H}{r}$ in a logarithmic scale. Note that $r = r_H$ then corresponds to $x = 0$ and $r = \infty$ corresponds to $x = 1$.

Whereas the metric and electromagnetic functions appear smooth at first glance, as seen in the inlets of Figures 2a,b and in Figure 2c, the scalar field function (Figure 2d) immediately reveals a critical behavior: As Q_{cr} is approached, the scalar field function assumes a finite limiting value $\Phi(r_H)$ at the imposed horizon $r_H = 2$. However, the function $\Phi(r)$ decreases more and more steeply as it approaches zero, with its boundary value at infinity, $\Phi(\infty) = 0$. In fact, in the limit $Q \to Q_{cr}$, it tends to zero already at a critical value of the radial coordinate $r = r_{cr}$, which coincides with the value of the critical charge, $r_{cr} = Q_{cr}$. In the limit, therefore, the solution features a finite scalar field in the interior $r < r_{cr}$, whereas the scalar field vanishes identically in the exterior $r > r_{cr}$.

Since the critical exterior solution is a pure electrovacuum solution, starting at r_{cr} and possessing electric charge Q_{cr}, this suggests that the exterior critical solution is described by an extremal RN black hole. However, instead of carrying charge $Q = 2$, it carries the critical charge Q_{cr}. Comparing the Q_{cr} numerical solution with a Q_{cr} extremal RN black hole shows that this conclusion holds true. In the interior, however, not only is the scalar field function finite but also all the functions differ from this Q_{cr} extremal RN black hole, as they must in order to satisfy the imposed boundary conditions at $r_H = 2$.

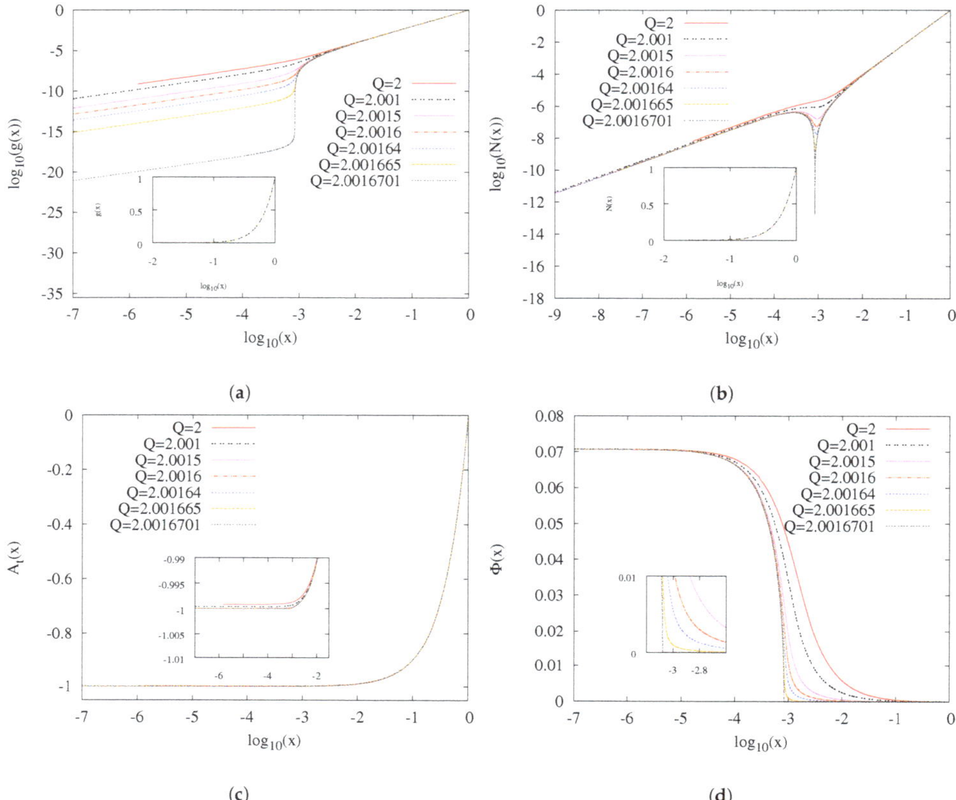

Figure 2. Approach to the critical solution for $\alpha = 200$: (**a**) metric function $g(r) = N(r)e^{-2\delta(r)}$, (**b**) metric function $N(r) = 1 - \frac{2m(r)}{r}$, (**c**) electromagnetic function $A_t(r)$ and (**d**) scalar function $\Phi(r)$ vs. the compactified radial coordinate $x = 1 - \frac{r_H}{r}$. The insets highlight the vicinity of the critical radius $r = r_{cr}$.

We now inspect the behavior of the functions in the interior in more detail as Q_{cr} is approached, starting with the electromagnetic function A_t. As seen in Figure 2c, also in the interior, a limiting critical solution is reached. At the critical radius r_{cr}, the electromagnetic function A_t of this critical solution assumes the value $A_t(r_{cr}) = -1$. In fact, in the full interior $r < r_{cr}$, it assumes this value $A_t(r \leq r_{cr}) = -1$, as seen in the inset of the figure. Therefore, there is no electric field in the interior region.

To reveal the critical behavior of the metric functions, we need to consider double logarithmic plots, as exhibited in Figure 2a,b. The extremal RN with charge Q_{cr} would have a double zero at r_{cr}, for both functions $g(x)$ and $N(r)$. Indeed, we observe a very sharp drop at r_{cr} as the limiting solution is approached for both metric functions $g(r)$ and $N(r)$, confirming our interpretation of the exterior solution. However, in the interior, it becomes apparent that we have not yet fully reached the critical solution but are only very close to it.

In the interior, both functions $g(r)$ and $N(r)$ differ distinctly. The function $N(r)$ tends to a finite limiting solution in the interior, except at $r_H = 2$, where the boundary conditions force it to vanish. In contrast to $N(r)$, the function $g(r)$ approaches zero in the limit. (With every further digit determined at the critical value Q_{cr}, the function $g(r)$ assumes smaller values in the interior.) Recalling that the function $g(r)$ has been decomposed into the factors $N(r)$ and $\exp(-2\delta(r))$, we conclude that it is the function $\delta(r)$ which causes $g(r)$ to vanish in the interior in the limit, since $N(r)$ has a finite limit. It is instructive now to look again at the horizon temperature T_H. Having observed above (e.g., in the inset of Figure 1b) that $T_H \to 0$ on the cold branch in the limit, one may expect that the reason was that a degenerate horizon arose at $r_H = 2$ and therefore the derivative $N'(r)$ vanished there. However, now, we see that there is no degenerate horizon at $r_H = 2$ in the limit. Instead, T_H vanishes because of the factor $\exp(-\delta(r))$ in Equation (18).

We have thus obtained the following scenario: In the limit $Q \to Q_{cr}$, the space time splits into an exterior and an interior part. The exterior is described by an extremal RN black hole; the interior has a finite scalar field but no electric field. Since at the critical radius r_{cr}, a double zero is approached, as featured by a degenerate horizon, the radial distance $l(r)$ to and from r_{cr} increases as the critical solution is approached. In the limit, both parts of the space time become infinite in extent.

To demonstrate this effect for the interior part, we consider in Figure 3 the radial distance $l(r)$, defined via

$$l(r) = \int_{r_H}^{r} \frac{dr}{\sqrt{N(r)}} \ . \tag{19}$$

The dependence of the radial coordinate $r(l)$ on the radial distance is illustrated in Figure 3a for the approach to the critical solution. For Q_{cr}, an infinite radial distance is reached at the finite value of the radial coordinate r_{cr}; thus, an infinite throat is formed. For $Q \to Q_{cr}$, the metric function $N(r)$ (Figure 3b), the electromagnetic function $A_t(r)$ (Figure 3c) and the scalar function $\Phi(r)$ (Figure 3d) are also shown versus the radial distance $l(r)$. The metric function $N(r)$ again highlights the formation of an infinite throat in the limit, while the electromagnetic function $A_t(r)$ approaches a constant value in the interior region. The scalar function $\Phi(r)$, on the other hand, demonstrates that, when considered as a function of the radial distance l instead of the radial coordinate r, there is remarkably little dependence on the value of the charge Q during the approach $Q \to Q_{cr}$. An analogous observation was made for the matter functions of the magnetic monopoles during their approach to their respective critical solution [46].

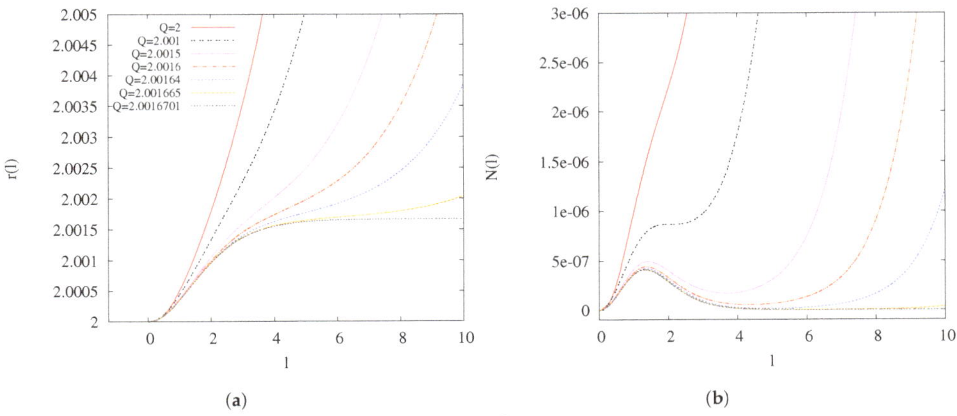

(a)   (b)

Figure 3. *Cont.*

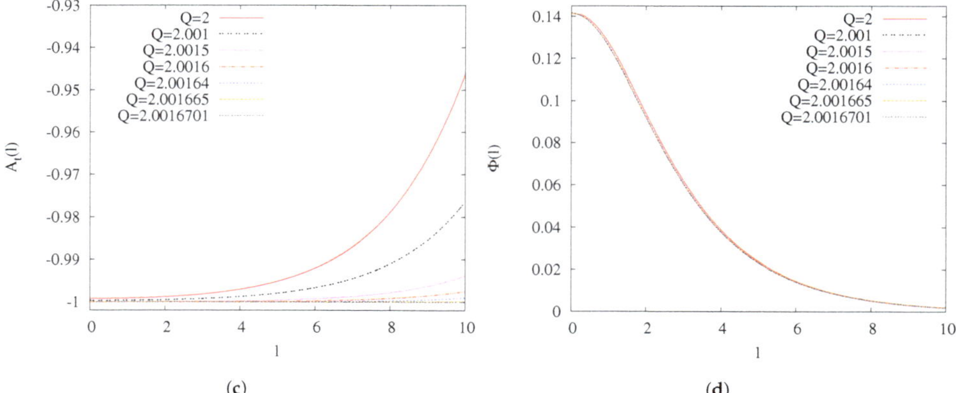

(c)
    (d)

Figure 3. Approach to the critical solution for $\alpha = 200$: (**a**) radial coordinate r, (**b**) metric function $N(r) = 1 - \frac{2m(r)}{r}$, (**c**) electromagnetic function $A_t(r)$ and (**d**) scalar function $\Phi(r)$ vs. the radial distance $l(r)$ (Equation (19)).

3.3. α-Dependence of the Critical Solution

We now demonstrate that the critical scenario described above holds for a large range of couplings α, showing that the scenario is rather generic. We exhibit the critical solutions in Figure 4, where we show the metric function $N(r) = 1 - \frac{2m(r)}{r}$ (Figure 4a) and the scalar function $\Phi(r)$ (Figure 4b) for a set of couplings α in the interval $[3.2, 200]$.

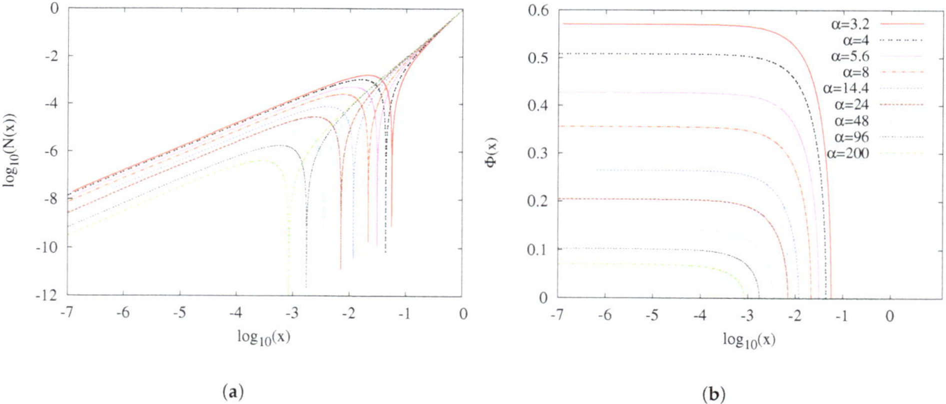

(a)
    (b)

Figure 4. Critical solutions for a set of couplings α: (**a**) metric function $N(r) = 1 - \frac{2m(r)}{r}$ and (**b**) scalar function $\Phi(r)$ vs. the compactified radial coordinate $x = 1 - \frac{r_H}{r}$. Note that the key applies to both figures.

The figure shows that, with decreasing α, the critical radius r_{cr} increases. We recall that it coincides with the critical charge, $r_{cr} = Q_{cr}$. At the same time, the scalar field assumes larger values in the interior. We highlight the α-dependence of the critical charge Q_{cr} and of the horizon value of the scalar field $\Phi(r_H)$ in Figure 5a,b, respectively. We note the steep increase of the critical charge Q_{cr} for small α, while $Q_{cr} \to r_H$ for large α. This dependence can be well described by the simple relation

$$\frac{Q_{cr}}{r_H} - 1 = \frac{1}{4\sqrt{2}\,\alpha}, \tag{20}$$

as demonstrated in the figure. The horizon value of the scalar field $\Phi(r_H)$ satisfies the even simpler relation

$$\Phi(r_H) = \frac{1}{\sqrt{\alpha}} \, , \tag{21}$$

as seen in the figure as well. In Figure 5c,d, we show that, for the interior critical solution, the derivatives of the metric function $m(r)$ and the scalar function $\Phi(r)$ at the horizon precisely respect the expansion at the horizon (Equations (12) and (13), respectively).

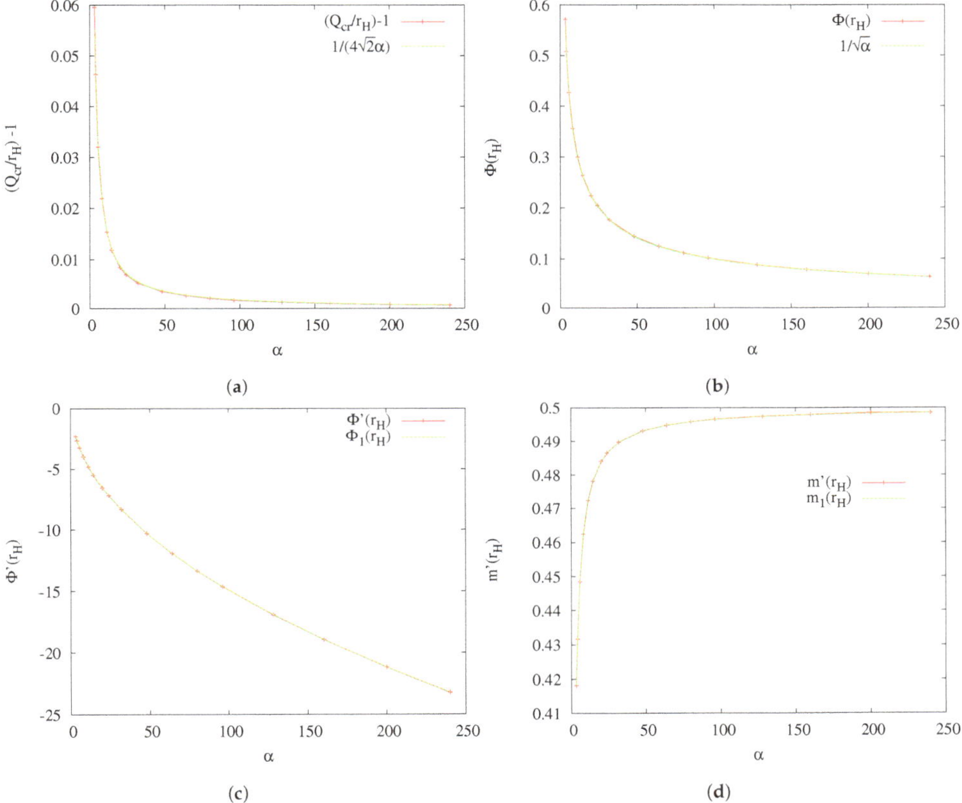

Figure 5. Properties of critical solutions: (**a**) $Q_{cr}(\alpha)$ (solid red) and limit for $\alpha \to \infty$ according to Equation (20) (dashed green), (**b**) $\Phi(r_H)(\alpha)$ (solid red) and relation from Equation (21) (dashed green), (**c**) $\Phi'(r_H)(\alpha)$ (solid red) and expansion coefficient $\Phi_1(r_H)(\alpha)$ from Equation (14) (dashed green), and (**d**) $m'(r_H)(\alpha)$ (solid red) and expansion coefficient $m_1(r_H)(\alpha)$ from Equation (14) (dashed green).

4. Excited Solutions

Until now, we have focused the discussion on the fundamental branch of scalarized black holes, with scalar field functions that possess no nodes. However, in certain cases, the model allows for the existence of excited solutions, meaning solutions with nodes in the scalar field functions. Let us discuss them briefly here.

These solutions present very similar properties to the $n = 0$ solutions, with a hot/cold branch structure, the hot one bifurcating from the cold one. The cold branch also reaches a critical end-point with $q = 1$. However, an important difference is that all of these excited branches are radially unstable, as previously discussed in [38,39]. Also, the number of excited branches depends on the coupling constant α, i.e., the larger the value of α, the more excited branches exist.

As an example, in Figure 6a, we show again the reduced area a_H as a function of the reduced charge q for $\alpha = 200$. This is the same as Figure 1b, but now, we include the $n = 1$ solutions in solid green. As we can see, these excited solutions could also be distinguished in two different branches: one branch (cold) would extend from $q = 1$ to a minimum $q_{min} < 1$; the second branch (hot) would extend from this q_{min} up to a certain $q_{max} > 1$. It turns out that the interval (q_{min}, q_{max}), where $n = 1$ solutions exist, is smaller than the domain interval of the $n = 0$ solutions. On the other hand, for this particular value of the coupling constant ($\alpha = 200$), only $n = 1$ excited solutions exist. However, sufficiently large values of α allow for additional branches with $n > 1$ excited solutions appear.

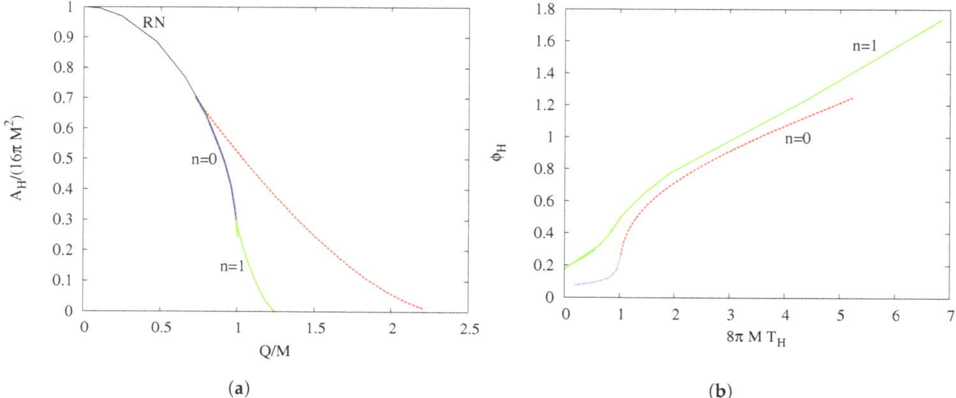

(a)   (b)

Figure 6. (**a**) Reduced horizon area a_H vs. q for $\alpha = 200$: In addition to the RN branch (solid black), the ($n = 0$) cold branch (finely dotted blue) and the ($n = 0$) hot branch (coarsely dotted red), the $n = 1$ excited solutions (solid green) are shown. (**b**) A similar figure for the scalar field at the horizon Φ_H vs. reduced temperature t_H.

In Figure 6b, we show the scalar field at the horizon as a function of the reduced temperature t_H, also for $\alpha = 200$. As we approach the $q = 1$ limit along the cold branch, the temperature goes to zero but the value of the scalar field remains constant. (This is similar to what happens for the $n = 0$ solutions we have discussed in the previous sections, as shown in this figure.) If we compare solutions with the same t_h, the larger the excitation number, the larger the value of the scalar field at the horizon.

As already said, the critical behavior is also present in these excited solutions. When fixing $r_H = 2$, the critical solutions possess a value of the electric charge $Q = Q_{cr}(n) > 2$ (this value depends on n). The limit results in a set of critical solutions with split space time, similar to the $n = 0$ solutions we have been discussing in the previous sections. The exterior part is the extremal RN solution. On the interior part, however, we have a nontrivial scalar field for which the number of nodes can be labeled by an integer number n.

In Figure 7, some profiles for $n = 1$ critical solutions with different values of α are shown as an example. In particular, we show the critical solutions for $\alpha = 200$ with $Q_{cr} = 2.02407591$ (solid red), for $\alpha = 400$ with $Q_{cr} = 2.01180482$ (dashed green) and for $\alpha = 800$ with $Q_{cr} = 2.00584391$ (dotted blue).

In Figure 7a, we show the metric function $N(r)$ versus $x = 1 - r_H/r$. It demonstrates that the behavior of these $n = 1$ solutions is very similar to the one presented in Figure 4a, which corresponds to $n = 0$ solutions. Essentially, the metric function $N(r)$ develops a zero at some critical value $r = r_{cr}$. For $r > r_{cr}$, the solution is extremal RN, but in the interior region, it is a nontrivial solution with a scalar field. The main difference now, as shown more clearly in the inset, is that the function develops a minimum at a certain point $r < r_{cr}$.

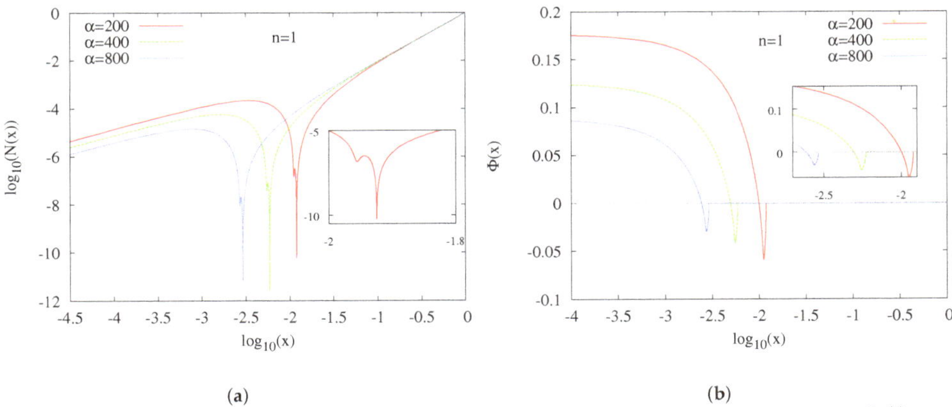

(a)

(b)

Figure 7. Critical solutions with $n = 1$ for a set of couplings α: (**a**) metric function $N(r) = 1 - \frac{2m(r)}{r}$ and (**b**) scalar function $\Phi(r)$ vs. the compactified radial coordinate $x = 1 - \frac{r_H}{r}$.

Figure 7b exhibits the scalar field Φ versus x, with the inset showing a zoomed region around the critical value of the radial coordinate r_{cr}. This demonstrates that the profile of the scalar field is quite different from the one in Figure 4b. Now the scalar field function not only vanishes at r_{cr} but also has a node at some intermediate point.

Solutions with more nodes also follow this pattern. As an example, we depict in Figure 8 similar plots for $n = 2$ solutions. These correspond to critical solutions for $\alpha = 640$ with $Q_{cr} = 2.0362514$ (solid red), for $\alpha = 832$ with $Q_{cr} = 2.0274443$ (dashed green) and for $\alpha = 960$ with $Q_{cr} = 2.02361315$ (dotted blue). Figure 8 (left) exhibits how the N function develops a minimum a bit below r_{cr}, and in Figure 8 (right), we demonstrate that the scalar field function has two nodes below r_{cr}.

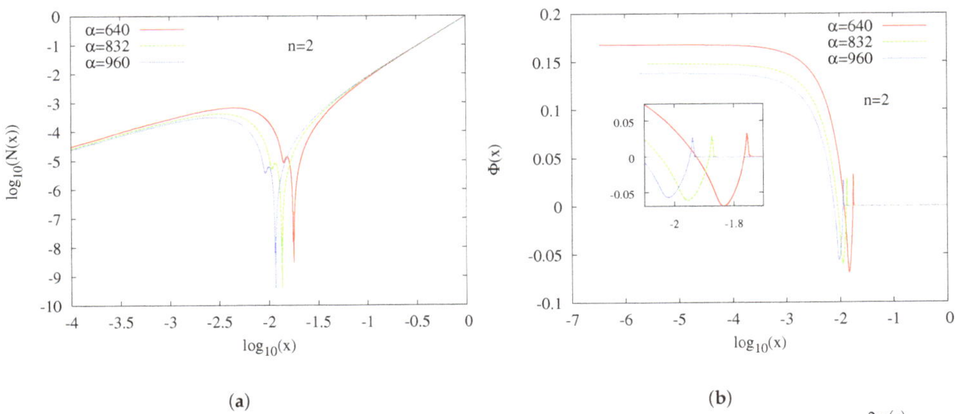

(a)

(b)

Figure 8. Critical solutions with $n = 2$ for a set of couplings α: (**a**) metric function $N(r) = 1 - \frac{2m(r)}{r}$ and (**b**) scalar function $\Phi(r)$ vs. the compactified radial coordinate $x = 1 - \frac{r_H}{r}$.

5. Conclusions

We have considered black holes in EMs models with a quartic coupling function $f(\Phi) = 1 + \alpha\Phi^4$, that allows for RN black holes as well as cold and hot scalarized black holes [38]. The models are parameterized by the coupling constant α and exhibit generic features. In particular, the RN black holes never become unstable to grow scalar hair [38,39]. Still, the cold branch follows the RN branch to a large extent and starts at $q = Q/M = 1$, the endpoint of the RN branch.

In particular, we have investigated the approach of the cold branch to this critical point $q = 1$ in detail for these EMs models. For that purpose, we have fixed the horizon radius r_H of the black holes and then increased the electromagnetic charge Q until, for some fixed coupling α, a critical solution with $q = 1$ has been reached. Whereas an extremal RN black hole with horizon radius r_H satisfies $q = 1$ with $r_H = Q = M$, the critical solution on the cold branch satisfies $q = 1$ with $r_{cr} = Q_{cr} = M_{cr}$ and $r_H < r_{cr}$. Consequently, the critical solution cannot simply be an extremal RN black hole.

Inspection of the functions of the critical solution then revealed that the critical solution splits the space time into two parts: an exterior part which indeed corresponds to an extremal RN black hole solution but with $r_{cr} = Q_{cr} = M_{cr}$, and an interior part, which has a finite scalar field that vanishes at r_{cr} and a vanishing electromagnetic field. Since the radial metric function develops a double zero at r_{cr}, both parts of the space time become infinite in size as the space time splits.

By studying the critical solution for many values of the coupling α, we have shown that this observed phenomenon is generic for these EMs models. In fact, the critical charge Q_{cr} possesses a rather simple α-dependence, with $(Q_{cr}/r_H - 1)$ being inversely proportional to α, while the horizon value of the scalar field Φ_H is inversely proportional the square root of α. At the same time, the functions of the critical solution satisfy the horizon expansion at r_H as well as the expansion at infinity, the latter of course with vanishing scalar charge.

In addition, we have discussed the case of the excited solutions, with nodes in the scalar field functions. These solutions posses a similar (hot/cold) branch structure, with a critical limit that splits the space time just like in the nodeless case. The excited solutions can be labeled by an integer number n, counting the number of nodes of the scalar in the interior part of the space time.

It is interesting that the critical phenomenon which has previously been encountered for non-Abelian magnetic monopoles and their associated hairy black holes as well as for similar non-Abelian solutions arises also in these EMs models. Note a somewhat similar observation in the recent investigation of strong gravity effects of charged Q-clouds and inflating black holes [54] (see also [55] in this issue). For non-Abelian monopoles, the interior solution has nontrivial non-Abelian gauge fields and Higgs field, whereas the exterior solution is simply an embedded RN black hole with magnetic charge. Not surprisingly, for solutions with both electric and magnetic charge, this phenomenon persists. Even pure non-Abelian solutions (without a Higgs field) were shown to exhibit a somewhat analogous phenomenon, which emerges in the limit of infinite node number (see e.g., [43]). It would thus be interesting to find general criteria to identify models that allow for this type of phenomenon where the space time splits into two infinitely extended parts, with the interior part containing nontrivial fields that vanish identically in the exterior part, where a simple extremal black hole solution emerges.

Author Contributions: Conceptualization, methodology, software, validation, formal analysis, investigation, resources, data curation, writing—original draft preparation, writing—review and editing, visualization, supervision, project administration, funding acquisition: S.K., J.K. and J.L.B.-S. All authors have read and agreed to the published version of the manuscript.

Funding: This research was funded by: DFG Research Training Group 1620 *Models of Gravity*; DFG Project No. BL 1553; COST Action CA15117; COST Action CA16104; FCT project PTDC/FISOUT/28407/2017; FCT project PTDC/FIS-AST/3041/2020

Acknowledgments: J.L.B.-S., S.K. and J.K. gratefully acknowledge support by the DFG Research Training Group 1620 *Models of Gravity*. J.L.B.-S. would like to acknowledge support from the DFG Project No. BL 1553 and the FCT projects PTDC/FISOUT/28407/2017 and PTDC/FIS-AST/3041/2020. We also would like to acknowledge networking support by the COST Actions CA15117 and CA16104.

Conflicts of Interest: The authors declare no conflict of interest.

References

1. Doneva, D.D.; Yazadjiev, S.S. New Gauss-Bonnet Black Holes with Curvature-Induced Scalarization in Extended Scalar-Tensor Theories. *Phys. Rev. Lett.* **2018**, *120*, 131103. [CrossRef]
2. Antoniou, G.; Bakopoulos, A.; Kanti, P. Evasion of No-Hair Theorems and Novel Black-Hole Solutions in Gauss-Bonnet Theories. *Phys. Rev. Lett.* **2018**, *120*, 131102. [CrossRef]
3. Silva, H.O.; Sakstein, J.; Gualtieri, L.; Sotiriou, T.P.; Berti, E. Spontaneous scalarization of black holes and compact stars from a Gauss-Bonnet coupling. *Phys. Rev. Lett.* **2018**, *120*, 131104. [CrossRef]
4. Antoniou, G.; Bakopoulos, A.; Kanti, P. Black-Hole Solutions with Scalar Hair in Einstein-Scalar-Gauss-Bonnet Theories. *Phys. Rev. D* **2018**, *97*, 084037. [CrossRef]
5. Blázquez-Salcedo, J.L.; Doneva, D.D.; Kunz, J.; Yazadjiev, S.S. Radial perturbations of the scalarized Einstein-Gauss-Bonnet black holes. *Phys. Rev. D* **2018**, *98*, 084011. [CrossRef]
6. Doneva, D.D.; Kiorpelidi, S.; Nedkova, P.G.; Papantonopoulos, E.; Yazadjiev, S.S. Charged Gauss-Bonnet black holes with curvature induced scalarization in the extended scalar-tensor theories. *Phys. Rev. D* **2018**, *98*, 104056. [CrossRef]
7. Minamitsuji, M.; Ikeda, T. Scalarized black holes in the presence of the coupling to Gauss-Bonnet gravity. *Phys. Rev. D* **2019**, *99*, 044017. [CrossRef]
8. Silva, H.O.; Macedo, C.F.; Sotiriou, T.P.; Gualtieri, L.; Sakstein, J.; Berti, E. Stability of scalarized black hole solutions in scalar-Gauss-Bonnet gravity. *Phys. Rev. D* **2019**, *99*, 064011. [CrossRef]
9. Brihaye, Y.; Ducobu, L. Hairy black holes, boson stars and non-minimal coupling to curvature invariants. *Phys. Lett. B* **2019**, *795*, 135–143. [CrossRef]
10. Doneva, D.D.; Staykov, K.V.; Yazadjiev, S.S. Gauss-Bonnet black holes with a massive scalar field. *Phys. Rev. D* **2019**, *99*, 104045. [CrossRef]
11. Myung, Y.S.; Zou, D.C. Black holes in Gauss–Bonnet and Chern–Simons-scalar theory. *Int. J. Mod. Phys. D* **2019**, *28*, 1950114. [CrossRef]
12. Cunha, P.V.; Herdeiro, C.A.; Radu, E. Spontaneously Scalarized Kerr Black Holes in Extended Scalar-Tensor-Gauss-Bonnet Gravity. *Phys. Rev. Lett.* **2019**, *123*, 011101. [CrossRef] [PubMed]
13. Macedo, C.F.; Sakstein, J.; Berti, E.; Gualtieri, L.; Silva, H.O.; Sotiriou, T.P. Self-interactions and Spontaneous Black Hole Scalarization. *Phys. Rev. D* **2019**, *99*, 104041. [CrossRef]
14. Hod, S. Spontaneous scalarization of Gauss-Bonnet black holes: Analytic treatment in the linearized regime. *Phys. Rev. D* **2019**, *100*, 064039. [CrossRef]
15. Collodel, L.G.; Kleihaus, B.; Kunz, J.; Berti, E. Spinning and excited black holes in Einstein-scalar-Gauss–Bonnet theory. *Class. Quantum Gravity* **2020**, *37*, 075018. [CrossRef]
16. Bakopoulos, A.; Kanti, P.; Pappas, N. Large and ultracompact Gauss-Bonnet black holes with a self-interacting scalar field. *Phys. Rev. D* **2020**, *101*, 084059. [CrossRef]
17. Blázquez-Salcedo, J.L.; Doneva, D.D.; Kahlen, S.; Kunz, J.; Nedkova, P.; Yazadjiev, S.S. Axial perturbations of the scalarized Einstein-Gauss-Bonnet black holes. *Phys. Rev. D* **2020**, *101*, 104006. [CrossRef]
18. Blázquez-Salcedo, J.L.; Doneva, D.D.; Kahlen, S.; Kunz, J.; Nedkova, P.; Yazadjiev, S.S. Polar quasinormal modes of the scalarized Einstein-Gauss-Bonnet black holes. *arXiv* **2020**; arXiv:2006.06006.
19. Dima, A.; Barausse, E.; Franchini, N.; Sotiriou, T.P. Spin-induced black hole spontaneous scalarization. *arXiv* **2020**, arXiv:2006.03095.
20. Doneva, D.D.; Collodel, L.G.; Krüger, C.J.; Yazadjiev, S.S. Black hole scalarization induced by the spin – 2+1 time evolution. *Phys. Rev. D* **2020**, *102*, 104027.
21. Berti, E.; Collodel, L.G.; Kleihaus, B.; Kunz, J. Spin-induced black-hole scalarization in Einstein-scalar-Gauss-Bonnet theory. *arXiv* **2020**, arXiv:2009.03905.
22. Herdeiro, C.A.; Radu, E.; Silva, H.O.; Sotiriou, T.P.; Yunes, N. Spin-induced scalarized black holes. *arXiv* **2020**, arXiv:2009.03904.
23. Herdeiro, C.A.; Radu, E.; Sanchis-Gual, N.; Font, J.A. Spontaneous Scalarization of Charged Black Holes. *Phys. Rev. Lett.* **2018**, *121*, 101102. [CrossRef] [PubMed]
24. Myung, Y.S.; Zou, D.C. Instability of Reissner–Nordström black hole in Einstein–Maxwell–scalar theory. *Eur. Phys. J. C* **2019**, *79*, 273. [CrossRef]
25. Boskovic, M.; Brito, R.; Cardoso, V.; Ikeda, T.; Witek, H. Axionic instabilities and new black hole solutions. *Phys. Rev. D* **2019**, *99*, 035006. [CrossRef]

26. Myung, Y.S.; Zou, D.C. Quasinormal modes of scalarized black holes in the Einstein–Maxwell–Scalar theory. *Phys. Lett. B* **2019**, *790*, 400–407. [CrossRef]

27. Fernandes, P.G.; Herdeiro, C.A.; Pombo, A.M.; Radu, E.; Sanchis-Gual, N. Spontaneous Scalarisation of Charged Black Holes: Coupling Dependence and Dynamical Features. *Class. Quantum Gravity* **2019**, *36*, 134002; Erratum: *Class. Quantum Gravity* **2020**, *37*, 049501. [CrossRef]

28. Brihaye, Y.; Hartmann, B. Spontaneous scalarization of charged black holes at the approach to extremality. *Phys. Lett. B* **2019**, *792*, 244–250. [CrossRef]

29. Herdeiro, C.A.; Oliveira, J.M. On the inexistence of solitons in Einstein–Maxwell–scalar models. *Class. Quantum Gravity* **2019**, *36*, 105015. [CrossRef]

30. Myung, Y.S.; Zou, D.C. Stability of scalarized charged black holes in the Einstein–Maxwell–scalar theory. *Eur. Phys. J. C* **2019**, *79*, 641. [CrossRef]

31. Astefanesei, D.; Herdeiro, C.; Pombo, A.; Radu, E. Einstein–Maxwell–scalar black holes: Classes of solutions, dyons and extremality. *JHEP* **2019**, *10*, 078. [CrossRef]

32. Konoplya, R.; Zhidenko, A. Analytical representation for metrics of scalarized Einstein-Maxwell black holes and their shadows. *Phys. Rev. D* **2019**, *100*, 044015. [CrossRef]

33. Fernandes, P.G.; Herdeiro, C.A.; Pombo, A.M.; Radu, E.; Sanchis-Gual, N. Charged black holes with axionic-type couplings: Classes of solutions and dynamical scalarization. *Phys. Rev. D* **2019**, *100*, 084045. [CrossRef]

34. Herdeiro, C.A.; Oliveira, J.M. On the inexistence of self-gravitating solitons in generalised axion electrodynamics. *Phys. Lett. B* **2020**, *800*, 135076. [CrossRef]

35. Zou, D.C.; Myung, Y.S. Scalarized charged black holes with scalar mass term. *Phys. Rev. D* **2019**, *100*, 124055. [CrossRef]

36. Brihaye, Y.; Herdeiro, C.; Radu, E. Black Hole Spontaneous Scalarisation with a Positive Cosmological Constant. *Phys. Lett. B* **2020**, *802*, 135269. [CrossRef]

37. Astefanesei, D.; Blázquez-Salcedo, J.L.; Herdeiro, C.; Radu, E.; Sanchis-Gual, N. Dynamically and thermodynamically stable black holes in Einstein-Maxwell-dilaton gravity. *JHEP* **2020**, *7*, 063. [CrossRef]

38. Blázquez-Salcedo, J.L.; Herdeiro, C.A.; Kunz, J.; Pombo, A.M.; Radu, E. Einstein–Maxwell–scalar black holes: The hot, the cold and the bald. *Phys. Lett. B* **2020**, *806*, 135493. [CrossRef]

39. Blázquez-Salcedo, J.L.; Herdeiro, C.A.; Kahlen, S.; Kunz, J.; Pombo, A.M.; Radu, E. Quasinormal modes of hot, cold and bald Einstein–Maxwell–scalar black holes. *arXiv* **2020**, arXiv:2008.11744.

40. Astefanesei, D.; Blázquez-Salcedo, J.L.; Gómez, F.; Rojas, R. Thermodynamically stable asymptotically flat hairy black holes with a dilaton potential: The general case. *arXiv* **2020**, arXiv:2009.01854.

41. Gibbons, G.; Maeda, K.i. Black Holes and Membranes in Higher Dimensional Theories with Dilaton Fields. *Nucl. Phys. B* **1988**, *298*, 741–775. [CrossRef]

42. Kanti, P.; Mavromatos, N.; Rizos, J.; Tamvakis, K.; Winstanley, E. Dilatonic black holes in higher curvature string gravity. *Phys. Rev. D* **1996**, *54*, 5049–5058. [CrossRef] [PubMed]

43. Volkov, M.S.; Gal'tsov, D.V. Gravitating nonAbelian solitons and black holes with Yang-Mills fields. *Phys. Rep.* **1999**, *319*, 1–83. [CrossRef]

44. Gal'tsov, D. Gravitating lumps. In Proceedings of the 16th International Conference on General Relativity and Gravitation (GR16), Durban, South Africa, 15–21 July 2001; pp. 142–161.10.1142/9789812776556_0006. [CrossRef]

45. Kleihaus, B.; Kunz, J.; Navarro-Lerida, F. Rotating black holes with non-Abelian hair. *Class. Quantum Gravity* **2016**, *33*, 234002. [CrossRef]

46. Lee, K.M.; Nair, V.; Weinberg, E.J. Black holes in magnetic monopoles. *Phys. Rev. D* **1992**, *45*, 2751–2761. [CrossRef] [PubMed]

47. Breitenlohner, P.; Forgacs, P.; Maison, D. Gravitating monopole solutions. *Nucl. Phys. B* **1992**, *383*, 357–376. [CrossRef]

48. Breitenlohner, P.; Forgacs, P.; Maison, D. Gravitating monopole solutions. 2. *Nucl. Phys. B* **1995**, *442*, 126–156. [CrossRef]

49. Ridgway, S.; Weinberg, E.J. Instabilities of magnetically charged black holes. *Phys. Rev. D* **1995**, *51*, 638–646. [CrossRef]

50. Brihaye, Y.; Hartmann, B.; Kunz, J. Gravitating dyons and dyonic black holes. *Phys. Lett. B* **1998**, *441*, 77–82. [CrossRef]

51. Hartmann, B.; Kleihaus, B.; Kunz, J. Gravitationally bound monopoles. *Phys. Rev. Lett.* **2001**, *86*, 1422–1425. [CrossRef]
52. Hartmann, B.; Kleihaus, B.; Kunz, J. Axially symmetric monopoles and black holes in Einstein-Yang-Mills-Higgs theory. *Phys. Rev. D* **2002**, *65*, 024027. [CrossRef]
53. Ascher, U.; Christiansen, J.; Russell, R. A Collocation Solver for Mixed Order Systems of Boundary Value Problems. *Math. Comput.* **1979**, *33*, 659–679. [CrossRef]
54. Brihaye, Y.; Hartmann, B. Strong gravity effects of charged Q-clouds and inflating black holes. *arXiv* **2020**, arXiv:2009.08293.
55. Brihaye, Y.; Cônsole, F.; Hartmann, B. Inflation inside non-topological defects and scalar black holes. *arXiv* **2020**, arXiv:2010.15625.

Article

Spontaneous Symmetry Breaking and Its Pattern of Scales

Maurizio Consoli [1] and Leonardo Cosmai [2,*]

[1] INFN—Sezione di Catania, Via S. Sofia 64, I-95129 Catania, Italy; maurizio.consoli@ct.infn.it
[2] INFN—Sezione di Bari, I-70126 Bari, Italy
*   Correspondence: leonardo.cosmai@ba.infn.it

Received: 12 November 2020; Accepted: 4 December 2020; Published: 9 December 2020

Abstract: Spontaneous Symmetry Breaking (SSB) in $\lambda\Phi^4$ theories is usually described as a 2nd-order phase transition. However, most recent lattice calculations indicate instead a weakly 1st-order phase transition as in the one-loop and Gaussian approximations to the effective potential. This modest change has non-trivial implications. In fact, in these schemes, the effective potential at the minima has two distinct mass scales: (i) a first mass m_h associated with its quadratic curvature and (ii) a second mass M_h associated with the zero-point energy which determines its depth. The two masses describe different momentum regions in the scalar propagator and turn out to be related by $M_h^2 \sim m_h^2 \ln(\Lambda_s/M_h)$, where Λ_s is the ultraviolet cutoff of the scalar sector. Our lattice simulations of the propagator are consistent with this two-mass picture and, in the Standard Model, point to a value $M_h \sim 700$ GeV. However, despite its rather large mass, this heavier excitation would interact with longitudinal W's and Z's with the same typical coupling of the lower-mass state and would therefore represent a rather narrow resonance. Two main novel implications are emphasized in this paper: (1) since vacuum stability depends on the much larger M_h, and not on m_h, SSB could originate within the pure scalar sector regardless of the other parameters of the theory (e.g., the vector-boson and top-quark mass) (2) if the smaller mass were fixed at the value $m_h = 125$ GeV measured at LHC, the hypothetical heavier state M_h would then naturally fit with the peak in the 4-lepton final state observed by the ATLAS Collaboration at 700 GeV.

Keywords: Spontaneous Symmetry Breaking; BEH field mass spectrum; LHC experiments

PACS: 11.30.Qc; 12.15.-y; 13.85.-t

---

1. Introduction

Spontaneous Symmetry Breaking (SSB) through the vacuum expectation value $\langle\Phi\rangle \neq 0$ of a fundamental scalar field, the BEH field [1,2], is an essential element of the Standard Model. This original idea has been recently confirmed by the discovery at LHC [3,4] of a narrow scalar resonance with mass $m_h \sim 125$ GeV whose characteristics fit well with the theoretical expectations. This has produced the widespread belief that any change of this general picture could only originate from new physics.

However, this conclusion might not be entirely true. In fact, at present, only the gauge and Yukawa interactions of the 125 GeV resonance have been tested. Instead, the possible effects of a genuine scalar self-coupling $\lambda = 3m_h^2/\langle\Phi\rangle^2$ are still below the precision of the observations. This suggests that some uncertainty on the origin of SSB may still persist.

Originally, the underlying mechanism was identified in a classical double-well, scalar potential. However, later, after Coleman and Weinberg [5], the classical potential was replaced by the quantum effective potential $V_{\text{eff}}(\varphi)$ which includes the zero-point energy of all fields in the theory.

Yet, SSB could still originate within the pure $\lambda\Phi^4$ sector if the other fields give a negligible contribution to the vacuum energy. To fully appreciate this point, we must start from scratch and consider one aspect which has still to be clarified: the nature of the phase transition in a pure $\lambda\Phi^4$ scalar theory in 4D. More precisely, is it a 2nd-order phase transition or a (weakly) 1st-order transition? Surprising as it may be, this apparently minor change can have substantial phenomenological implications.

To this end, in Sections 2–4 we will give a general overview of the problem and argue that SSB in pure $\lambda\Phi^4$ theory is a weak 1st-order phase transition. Then, in this picture, besides the known resonance with mass $m_h \sim 125$ GeV, we expect a new excitation of the BEH field with a much larger mass $M_h \sim 700$ GeV. Since vacuum stability depends on this larger M_h, and not on m_h, SSB could well originate within the pure scalar sector regardless of the remaining parameters of the theory (as the vector boson or top-quark mass).

However, despite such large mass, this heavier state would interact with longitudinal W's and Z's with the same typical strength of the lower-mass state. As such, it would represent a rather narrow resonance. On this basis, in Sections 5 and 6, we will consider these more phenomenological aspects and their implications for the present LHC experiments.

2. SSB: 2nd- or (Weak) 1st-Order Phase Transition?

To introduce the problem, let us start with the classical potential ($\lambda > 0$)

$$V_{\text{class}}(\varphi) = \frac{1}{2}m^2\varphi^2 + \frac{\lambda}{4!}\varphi^4 \tag{1}$$

Here, there is no ambiguity. As one varies the m^2 parameter, one finds a 2nd-order phase transition occurring for $m^2 = 0$. However, in the full quantum theory is this conclusion still so obvious? To this end, one should look at the effective potential and study vacuum stability depending on the physical mass, say m_Φ^2, in the symmetric vacuum at $\varphi = 0$

$$V_{\text{eff}}''(\varphi = 0) \equiv m_\Phi^2 \tag{2}$$

Clearly, this is locally stable if $m_\Phi^2 > 0$. However, for $m_\Phi^2 > 0$, is this symmetric vacuum also *globally* stable? Or, instead, could the SSB transition be 1st-order and occur for some very small but still positive $m_\Phi^2 = m_c^2 > 0$? Then, if this were true, the lowest-energy state for the classically scale-invariant case $m_\Phi^2 = 0$ would correspond to the broken-symmetry phase with an expectation $\langle\Phi\rangle \neq 0$.

This dilemma, on the nature of the phase transition, goes back to the pioneering work of Coleman and Weinberg [5]. After subtracting a $\varphi-$ independent constant and quadratic divergences, in this massless limit of $\lambda\Phi^4$, their original 1-loop result was

$$V_{1-\text{loop}}(\varphi) = \frac{\lambda}{4!}\varphi^4 + \frac{\lambda^2\varphi^4}{256\pi^2}\left[\ln(\tfrac{1}{2}\lambda\varphi^2/\Lambda_s^2) - \frac{1}{2}\right] \tag{3}$$

where Λ_s is a large ultraviolet cutoff. As it is well known, this 1-loop form could equivalently be expressed as the sum of classical background + zero-point energy of a field with a $\varphi-$dependent mass $M(\varphi)$ given by

$$M^2(\varphi) \equiv \frac{1}{2}\lambda\varphi^2 \tag{4}$$

namely

$$V_{1-\text{loop}}(\varphi) = \frac{\lambda\varphi^4}{4!} - \frac{M^4(\varphi)}{64\pi^2}\ln\frac{\Lambda_s^2\sqrt{e}}{M^2(\varphi)} \tag{5}$$

By using this notation, there are non-trivial minima for those values, say $\varphi = \pm v$, where

$$M_h^2 \equiv M^2(\pm v) = \frac{\lambda v^2}{2} = \Lambda_s^2 \exp(-\frac{32\pi^2}{3\lambda}) \tag{6}$$

Therefore, since the massless theory exhibits SSB, the 1-loop potential indicates a 1st-order phase transition. Actually, it is a *weak* 1st-order transition because, in units of the M_h^2 in Equation (6), the mass m_Φ in the symmetric phase is bounded to be smaller than a critical mass [6]

$$m_\Phi^2 < m_c^2 = \frac{\lambda M_h^2}{64\pi^2\sqrt{e}} \sim \frac{M_h^2}{\ln(\Lambda_s/M_h)} \ll M_h^2 \tag{7}$$

With such extremely small critical mass, SSB emerges as an infinitesimally weak 1st-order transition which could hardly be distinguished from a 2nd-order transition unless one looks on an extremely fine scale.

As is well known [5], though, the standard Renormalization-Group (RG) improvement of the 1-loop potential contradicts this scenario. Indeed, leading-logarithmic terms entering the effective potential are re-absorbed into an effective coupling $\lambda(\varphi)$ giving a re-summed expression

$$V_{\mathrm{RG}}(\varphi) \sim \frac{\lambda(\varphi)}{4!}\varphi^4 \tag{8}$$

Thus, by restricting to $\lambda(\varphi) > 0$, the 1-loop minimum disappears and we would again predict a 2nd-order transition at $m_\Phi^2 = 0$. The standard view is that it is this latter point of view to be reliable.

To see why things are not so simple, let us consider another approximation scheme. Specifically, the Gaussian effective potential [7,8]. Diagrammatically, this corresponds to the infinite re-summation of all one-loop bubbles with mass $M(\varphi)$ and has a variational nature by exploring the Hamiltonian operator within the Gaussian functional states. For this reason, it is a very natural alternative because a Gaussian set of Green's functions would fit with the "triviality" of $\lambda\Phi^4$ theory in 4D. An early calculation [9] of the Gaussian effective potential for the one-component $\lambda\Phi^4$ theory confirmed the 1st-order scenario in agreement with the 1-loop potential. This is because the existing corrections beyond 1-loop reproduce the some functional form and thus support the same 1st-order picture.

Further calculations, by Bryhaye and one of us [10,11], confirmed that by imposing $V''_{\mathrm{Gauss}}(\varphi = 0) = 0$, the Gaussian effective potential for the O(2) and O(N)-symmetric scalar theories exhibits SSB thus again supporting the weak 1st-order picture. In particular, it was noted the non-uniformity of the two limits $N \to \infty$ and ultraviolet cutoff $\Lambda_s \to \infty$.

To fully appreciate the substantial equivalence with the one-loop potential, we observe that the infinite additional terms in the Gaussian effective potential can be expressed in a form analogous to Equation (5) with a simple redefinition of the classical background and of the $\varphi-$dependent mass in the zero-point energy, i.e.,

$$V_{\mathrm{Gauss}}(\varphi) = \frac{\hat{\lambda}\varphi^4}{4!} - \frac{\Omega^4(\varphi)}{64\pi^2}\ln\frac{\Lambda_s^2\sqrt{e}}{\Omega^2(\varphi)} \tag{9}$$

with

$$\hat{\lambda} = \frac{\lambda}{1 + \frac{\lambda}{16\pi^2}\ln\frac{\Lambda_s}{\Omega(\varphi)}} \qquad \text{and} \qquad \Omega^2(\varphi) = \frac{\hat{\lambda}\varphi^2}{2} \tag{10}$$

This shows that the 1-loop potential also admits a non-perturbative interpretation. In fact, by displaying the same basic structure of classical background + zero-point energy, it represents the prototype of all gaussian and post-gaussian calculations [12,13]. At the same time, it also explains why 1-loop and Gaussian approximations, although differing in terms of the bare parameters, can become identical in a suitable renormalization scheme [14,15].

This concordance among various approximations may cast some doubts on the re-summation in Equation (8) and its 2nd-order scenario. Nevertheless, at the time of those works, the precise motivation for the discrepancy was not understood. Thus, the whole problem of SSB in pure $\lambda\Phi^4$ theories did not attract much attention, also due to the lack of definite phenomenological implications.

However, two subsequent theoretical developments, producing new evidence in favor of the 1st-order scenario, have refreshed anew the interest into the whole problem:

(i) the first development was concerning the physical mechanisms [6] underlying SSB as a 1st-order transition. In fact, once SSB really coexists with a physical mass $0 < m_\Phi^2 \leq m_c^2$ for the elementary quanta of the symmetric phase, these quanta, the "phions" [6], should be considered to be real particles although, being "frozen" in the broken-symmetry vacuum, they would not be directly observable (like quarks). Now, the conventional picture of $\lambda\Phi^4$ corresponds to a repulsive interaction. Only its strength decreases at large distance. However, then, this is somewhat mysterious. In fact, if the interaction remains always repulsive, how could this broken-symmetry vacuum with $\langle\Phi\rangle \neq 0$, a Bose condensate of phions, have a lower energy than the $\langle\Phi\rangle = 0$ empty state with no phions? Here, a crucial observation [6] was that phions, moreover the $+\lambda\delta^3(\mathbf{r})$ contact repulsion, also feel a $-\lambda^2 \frac{e^{-2m_\Phi r}}{r^3}$ attraction arising at 1-loop and which becomes more and more important when $m_\Phi \to 0$ (From the scattering amplitude $\mathcal{M}$, computed from Feynman graphs, one can define an interparticle potential which is nothing but the 3D Fourier transform of $\mathcal{M}$, see Feinberg et al. [16,17].). By including both effects, one can now understand [6] why, for small enough m_Φ, the attraction can dominate and the lowest-energy state becomes a state with a non-zero density of phions Bose-condensed in the zero-momentum state.

However, then, if SSB is produced by these two competing effects (short-range repulsion and long-range attraction) we now understand the failure of the standard RG-analysis. In fact, the attractive term originates from the *ultraviolet-finite* part of the 1-loop graphs. Therefore, to correctly include higher-order effects, one should renormalize *both* the tree-level contact repulsion and the 1-loop, long-range attraction, as if there were *two* different coupling constants in the theory. This different procedure has been adopted by Stevenson [18], see Figure 1. By avoiding double counting, he has shown that the simple 1-loop result and its RG-improvement, in this new scheme, now agree very well so that the weak 1st-order scenario is confirmed.

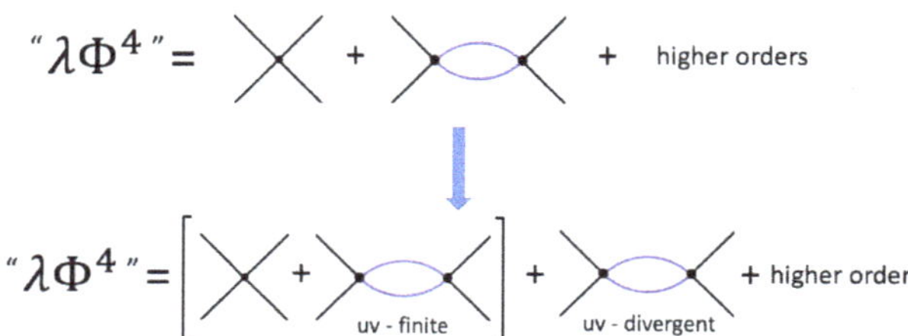

Figure 1. The re-arrangement of perturbation theory introduced by Stevenson [18] in his alternative analysis of $V_{\text{eff}}(\varphi)$. The quanta of the symmetric phase with mass m_Φ, besides the contact $+\lambda\delta^3(\mathbf{r})$ repulsion, also feel a $-\lambda^2 \frac{e^{-2m_\Phi r}}{r^3}$ attraction from the Fourier transform of the ultraviolet-finite part of the 1-loop term [6]. Its range diverges in the $m_\Phi \to 0$ limit and, for m_Φ below a critical mass m_c, the attraction will dominate and induce SSB. Since higher-order contributions simply renormalize these two basic effects, the resulting RG-improvement, in this new scheme, now confirms the 1st-order phase transition scenario as at 1-loop.

(ii) recent lattice simulations of pure $\lambda\Phi^4$ in 4D [19–21], obtained with different algorithms in the Ising limit of the theory (and on the present largest available lattices), indicate that the SSB phase transition is weakly 1st-order.

Since the above arguments (i) and (ii) confirm the 1st-order picture of SSB, and the general validity of the 1-loop and Gaussian approximations to the effective potential, we will now consider in Section 3 some important physical implications of this scenario.

3. Two-Mass Scales in the Broken Phase

To explore the physical implications of a 1st-order scenario of SSB, we will restrict to the one-loop approximation Equation (5) of $V_{\text{eff}}(\varphi)$ which is equivalent to the Gaussian approximation result

Equation (9). Equation (5) is just a different way of re-writing Equation (3) but intuitively supports the traditional view where the broken-symmetry phase is a simple massive theory with mass M_h as in Equation (6). Thus, one expects that up to small perturbative corrections, this is the mass parameter entering the scalar propagator.

To see why, again, things are not so simple, let us compute the quadratic shape of the effective potential, i.e., its second derivative at the minimum. This other quantity, say m_h^2, has the value

$$m_h^2 \equiv V_{\text{eff}}''(\pm v) = \frac{\lambda^2 v^2}{32\pi^2} = \frac{\lambda}{16\pi^2} M_h^2 \sim \frac{M_h^2}{L} \ll M_h^2 \tag{11}$$

where $L \equiv \ln \frac{\Lambda_s}{M_h}$. Now, the derivatives of the effective potential are just (minus) the n-point functions for zero external momentum. In particular, one finds

$$m_h^2 \equiv V_{\text{eff}}''(\varphi = \pm v) = -\Pi(p=0) = |\Pi(p=0)| \tag{12}$$

Therefore, by expressing the inverse propagator as

$$G^{-1}(p) = p^2 - \Pi(p) \tag{13}$$

we find $G^{-1}(p) \sim (p^2 + m_h^2)$ for $p \to 0$. This means that apparently, it is this smaller mass m_h^2, and not M_h^2, which enters the (low-momentum) propagator. However, now, in the $\lambda \to 0$ limit, m_h^2 and M_h^2 are vastly different scales (i.e., do not differ by small perturbative corrections). Thus one may ask: which is the right mass?

To better understand this point, let us sharpen the meaning of M_h by using the general relation which expresses the zero-point-energy ("zpe") in terms of the trace of the logarithm of $G^{-1}(p)$, i.e.,

$$zpe = \frac{1}{2} \int \frac{d^4p}{(2\pi)^4} \ln(p^2 - \Pi(p)) \tag{14}$$

Thus, after subtracting a constant and quadratic divergences, to match the 1-loop Equation (5), we can impose appropriate limits in the logarithmic divergent part (i.e., $p_{\text{max}}^2 \sim \sqrt{e}\Lambda_s^2$ and $p_{\text{min}}^2 \sim M_h^2$)

$$zpe = -\frac{1}{4} \int_{p_{\text{min}}}^{p_{\text{max}}} \frac{d^4p}{(2\pi)^4} \frac{\Pi^2(p)}{p^4} \sim -\frac{\langle \Pi^2(p) \rangle}{64\pi^2} \ln \frac{p_{\text{max}}^2}{p_{\text{min}}^2} \sim -\frac{M_h^4}{64\pi^2} \ln \frac{\sqrt{e}\Lambda_s^2}{M_h^2} \tag{15}$$

This relation indicates that M_h^4 reflects the typical, average $\langle \Pi^2(p) \rangle$ at non-zero p^2. Therefore, if we trust in the 1-loop relation $M_h^2 \sim m_h^2 \ln \frac{\Lambda_s}{M_h}$, we should observe large deviations in the propagator if we try to extrapolate to higher-p^2 with the 1-particle form $G^{-1}(p) \sim (p^2 + m_h^2)$ which is valid for $p \to 0$. In other words, in a 1st-order picture of SSB, the idea of a simple massive propagator seems to be wrong.

To show that these are not just speculations, let us compare with lattice calculations of the scalar propagator in the broken-symmetry phase. The simulation was performed [22] in the 4D Ising limit which has always been considered a convenient laboratory to exploit the non-perturbative aspects of the theory. It is the $\lambda\Phi^4$ in the limit of an infinite bare coupling $\lambda_0 = +\infty$, as sitting exactly at the Landau pole. As such, for a finite cutoff Λ_s, it represents the best possible definition of the local limit for a non-zero, low-energy coupling $\lambda \sim 1/L$ (where $L = \ln(\Lambda_s/M_h)$). For the convenience of the reader, we will report here the main results of [22].

In the Ising limit, the broken-symmetry phase corresponds to values of the basic hopping parameter $\kappa > \kappa_c$, with the critical $\kappa_c = 0.0748474(3)$ [19,20]. We computed the field vacuum expectation value

$$v = \langle |\phi| \rangle \quad , \quad \phi \equiv \frac{1}{V_4} \sum_x \phi(x) \tag{16}$$

and the connected propagator

$$G(x) = \langle \phi(x)\phi(0) \rangle - v^2 \tag{17}$$

where with $\langle ... \rangle$ we are indicating the average over lattice configurations.

In terms of the Fourier transform of the propagator, the extraction of m_h is straightforward, i.e.,

$$G(p = 0) = \frac{1}{|\Pi(p = 0)|} \equiv \frac{1}{m_h^2} \tag{18}$$

Instead M_h had to be extracted from the data for the Fourier transformed propagator at higher momentum. To this end, we first fitted the data to the 2-parameter form

$$G_{\text{fit}}(p) = \frac{Z_{\text{prop}}}{\hat{p}^2 + m_{\text{latt}}^2} \tag{19}$$

in terms of the lattice squared momentum $\hat{p}^2$ with $\hat{p}_\mu = 2 \sin p_\mu/2$. The quality of this fit was then studied to understand how reliable the determination $M_h \equiv m_{\text{latt}}$ is from the higher-momentum region. Finally, the propagator data were re-scaled by the factor $(\hat{p}^2 + m_{\text{latt}}^2)$. In this way, deviations from a straight line will show up clearly if a fitted mass $M_h \equiv m_{\text{latt}}$ fails to describe the lattice data when $p \to 0$.

The results in the symmetric phase, see Figure 2, show that there, with just a single lattice mass one can describe all data down to $p = 0$.

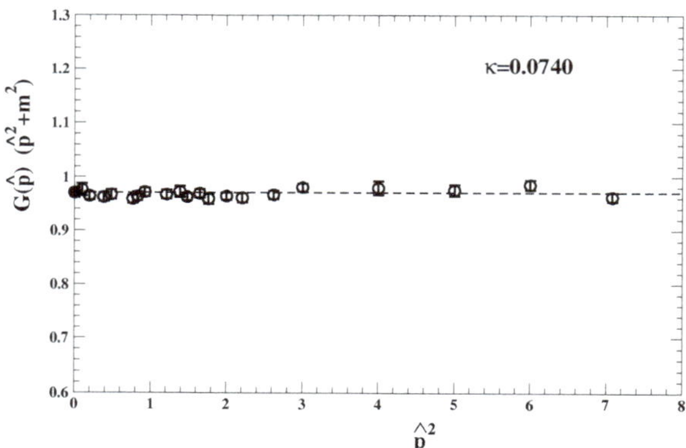

Figure 2. The data for the re-scaled lattice propagator ref. [22] in the symmetric phase at $\kappa = 0.074$ depending on the square lattice momentum $\hat{p}^2$ with $\hat{p}_\mu = 2 \sin p_\mu/2$. In this case, the mass fitted from higher-$\hat{p}^2$, $M_h = m_{\text{latt}} = 0.2141(28)$, describes well the data down to $p = 0$. The dashed line is the fitted $Z_{\text{prop}} = 0.9682(23)$.

In the broken phase, for $\kappa = 0.0749$, the results for the largest lattice 76^4 are reported in Figures 3 and 4. The larger mass obtained from the higher-momentum fit $\hat{p}^2 > 0.1$ was $M_h \equiv m_{\text{latt}} = 0.0933(28)$. As one can see from Figure 3, this fitted mass describes the data for not too small momentum. But for $p \to 0$ the deviations from a straight line become highly significant statistically. In this low-$\hat{p}^2$ limit, in fact, the data would require the other mass $m_h = |\Pi(p = 0)|^{1/2} = 0.0769$, see Figure 4.

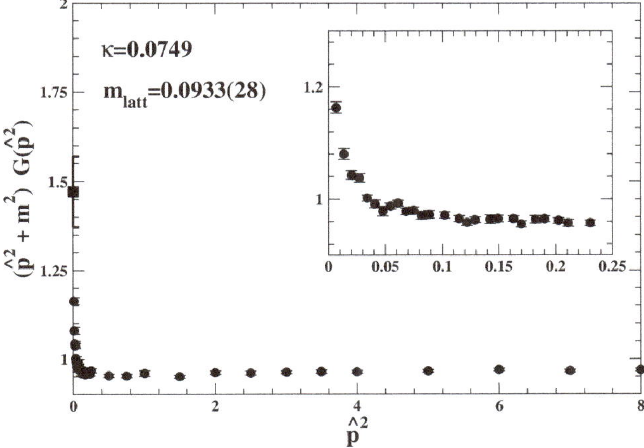

Figure 3. The data for the re-scaled lattice propagator ref. [22] in the broken phase at $\kappa = 0.0749$. The mass used for the re-scaling, $M_h = m_{\text{latt}} = 0.0933(28)$, was obtained from fitting to all data with $\hat{p}^2 > 0.1$. The black square at $p = 0$ is $Z(p = 0) = M_h^2/m_h^2 = 1.47(9)$ as computed from the fitted M_h for $m_h = |\Pi(p = 0)|^{1/2} = 0.0769$.

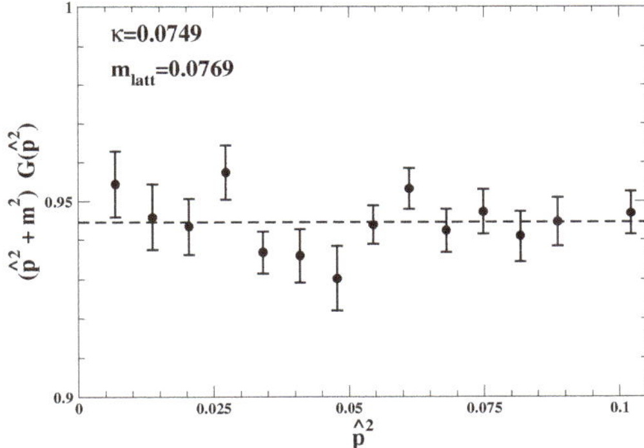

Figure 4. The lattice data of ref. [22] at $\kappa = 0.0749$ for $\hat{p}^2 < 0.1$. The mass used here for the re-scaling has been fixed at the value $m_h = |\Pi(p = 0)|^{1/2} = 0.0769$.

The difference between $M_h = 0.0933(28)$ and $m_h = 0.0769$ has the high statistical significance of 6 sigma. More importantly, once m_h^2 is directly computed from the zero-momentum limit of $G(p)$ and M_h is extracted from its behavior at higher p^2, the extrapolation of the results toward the critical point [22] is well consistent with the expected increasing logarithmic trend $M_h^2 \sim L m_h^2$.

4. The Relative Magnitude of m_h, M_h and $\langle \Phi \rangle$

As summarized in Section 3, our lattice simulations supports the idea of a scalar propagator which, in the broken phase, interpolates between two different mass scales m_h and M_h (Two-mass scales also require some interpolating form for the scalar propagator in loop corrections. Since some precise measurements, e.g., A_{FB} of the b-quark or $\sin^2 \theta_w$ from NC experiments [23], still favor a rather large BEH particle mass, this could help to improve the present rather low quality of the overall Standard Model fit). The lattice data are also consistent with the trend $M_h^2 \sim m_h^2 \ln(\Lambda_s/M_h)$ predicted by the one-loop and Gaussian approximations to the effective potential. Since the two masses do not scale uniformly in the $\Lambda_s \to \infty$ limit (This non- uniform scaling is crucial not to run in contradiction with the "triviality" of $\lambda \Phi^4$ in 4D [22]. In fact, this implies a continuum limit with a Gaussian set of Green's functions and therefore with a massive free-field propagator. Thus, in an ideal continuum theory, there can only be one mass depending on the unit of mass (m_h or M_h) adopted for measuring momenta), the question naturally arises about the extension to the Standard Model and their relationship with the fundamental weak scale $\langle \Phi \rangle \sim (G_{\text{Fermi}} \sqrt{2})^{-1/2} \sim 246.2$ GeV. In fact, it seems that we should now introduce two different coupling constants, say $m_h^2/\langle \Phi \rangle^2$ and $M_h^2/\langle \Phi \rangle^2$. However, then, since $M_h^2 \sim L m_h^2 \gg m_h^2$, are we faced with a weak- or a strong-coupling theory?

To approach the problem in a systematic way, let us first return to the one-loop relations Equations (5) and (6) in Section 2 and observe that the vacuum energy depends on M_h, *not* on m_h, namely

$$\mathcal{E} = V_{\text{eff}}(\pm v) = -\frac{M_h^4}{128\pi^2} = \text{const. } \Lambda_s^4 \exp(-\frac{64\pi^2}{3\lambda}) \tag{20}$$

This means that the critical temperature to restore the symmetry, $k_B T_c \sim M_h$, and the whole stability of the broken-symmetry phase will depend on M_h, not on m_h.

This remark will be crucial to understand the cutoff dependence of the various scales and to formulate a description of SSB which in principle can be extended to the $\Lambda_s \to \infty$ limit. In fact, since for any non-zero low-energy coupling λ there is a Landau pole Λ_s, we will consider the entire set of pairs (Λ_s, λ), (Λ_s', λ'), (Λ_s'', λ'')...with larger and larger cutoffs, smaller and smaller couplings but all with the same vacuum energy as in Equation (20). This amounts to impose

$$\left(\Lambda_s \frac{\partial}{\partial \Lambda_s} + \Lambda_s \frac{\partial \lambda}{\partial \Lambda_s} \frac{\partial}{\partial \lambda} \right) \mathcal{E}(\lambda, \Lambda_s) = 0 \tag{21}$$

a condition which can be derived from the more general requirement of RG-invariance for the effective potential in the $(\varphi, \lambda, \Lambda_s)$ 3-space

$$\left(\Lambda_s \frac{\partial}{\partial \Lambda_s} + \Lambda_s \frac{\partial \lambda}{\partial \Lambda_s} \frac{\partial}{\partial \lambda} + \Lambda_s \frac{\partial \varphi}{\partial \Lambda_s} \frac{\partial}{\partial \varphi} \right) V_{\text{eff}}(\varphi, \lambda, \Lambda_s) = 0 \tag{22}$$

In fact, for $\varphi = \pm v$, where $(\partial V_{\text{eff}}/\partial \varphi) = 0$, Equation (21) follows directly from (22).

It is important that in this RG-analysis, besides a first invariant mass scale $\mathcal{I}_1 = M_h$, if we introduce an anomalous dimension for the vacuum field

$$\Lambda_s \frac{\partial \varphi}{\partial \Lambda_s} \equiv \gamma(\lambda)\varphi \tag{23}$$

there will be a second invariant [22] associated with the RG-evolution in the $(\varphi, \lambda, \Lambda_s)$ 3-space, namely

$$\mathcal{I}_2(\varphi) = \varphi \exp(\int^\lambda dx \frac{\gamma(x)}{\beta(x)}) \tag{24}$$

This invariant fixes a particular normalization (The anomalous dimension of φ reflects the fact that from Equation (6), the cutoff-independent combination is $\lambda v^2 \sim M_h^2 = \mathcal{I}_1^2$ and not v^2 itself implying $\gamma = \beta/(2\lambda)$ [22]. This somewhat resembles the definition of the physical gluon condensate in QCD which is $\langle g^2 F_{\mu\nu}^a F^{a\mu\nu} \rangle$ and not just $\langle F_{\mu\nu}^a F^{a\mu\nu} \rangle$.) of φ and is then the natural candidate to represent the weak scale $\mathcal{I}_2(v) = \langle \Phi \rangle \sim 246.2$ GeV. The minimization of the effective potential is then translated into a proportionality of the two invariants through some constant K, say

$$M_h = K\langle \Phi \rangle \tag{25}$$

Such guiding principle indicates that M_h and $\langle \Phi \rangle$ scale uniformly while at the same time, $M_h^2 \sim L m_h^2$ and $\langle \Phi \rangle^2 \sim L m_h^2$. Therefore, by assuming the theoretical predictions for the ratio $m_h/\langle \Phi \rangle$, and computing the M_h/m_h ratio from our lattice data for the propagator, we have extracted the constant K. As shown in [22] such procedure, where the cutoff-dependent L drops out, leads to a final estimate $K = 2.92 \pm 0.12$ or

$$M_h \sim 720 \pm 30 \,\text{GeV} \qquad (26)$$

which includes various statistical and theoretical uncertainties and updates the previous work of refs. [24,25].

We emphasize that the relation $M_h = K\langle\Phi\rangle$ does not introduce a new large coupling $3K^2 = O(10)$ which modifies the phenomenology of the broken phase. This $3K^2$ is clearly quite distinct from the other coupling $\lambda = 3m_h^2/\langle\Phi\rangle^2 \sim 1/L$ but should not be viewed as a coupling producing *observable* interactions. Since M_h^4 reflects the magnitude of the vacuum energy density, it would be natural to consider $K^2 \sim \lambda L$ as a *collective* self-interaction of the vacuum condensate which persists when $\Lambda_s \to \infty$. This original view [14,15] can intuitively be formulated in terms of a scalar condensate whose increasing density $\sim L$ [6] compensates for the decreasing strength $\lambda \sim 1/L$ of the two-body coupling (This view of SSB has some analogy with the occurring of superconductivity in solid-state physics. There, the superconductive phase occurs even for an arbitrary small two-body attraction ϵ between the two electrons in a Cooper pair. However, the energy density and the collective quantities of the superconductive phase (as energy gap, critical temperature, etc.) depend on a much larger coupling ϵN obtained by re-scaling ϵ with the large density of states at the Fermi surface. This means that the same macroscopic description could be obtained with smaller and smaller ϵ and Fermi systems with suitably larger and larger N. In this analogy λ is the counterpart of ϵ and K^2 of ϵN).

Instead, $\lambda \sim 1/L$ is the right coupling for the *individual* interactions of the vacuum excitations, i.e., the BEH field and the Goldstone bosons. Consistently with the "triviality" of $\lambda\Phi^4$ theory, these interactions will become weaker and weaker when $\Lambda_s \to \infty$.

With this description of the scalar sector, and by using the Equivalence Theorem [26,27], the same conclusion applies to the high-energy interactions of the BEH field with the longitudinal vector bosons in the full $g_{\text{gauge}} \neq 0$ theory. In fact, the limit of zero-gauge coupling is smooth [28]. Therefore, up to corrections proportional to g_{gauge}, a heavy BEH resonance will interact exactly with the same strength as in the $g_{\text{gauge}} = 0$ theory [29]. For the convenience of the reader, this point will be summarized in Section 5. In Section 6, we will instead consider some phenomenological implications for the present LHC experiments.

5. Observable Interactions for a Large M_h

As anticipated, the quantity $3K^2$ should be understood as a collective self-coupling of the scalar condensate whose effects are re-absorbed into the vacuum structure. As such, it is basically different from the coupling λ defined through the $\beta-$function

$$\ln \frac{\mu}{\Lambda_s} = \int_{\lambda_0}^{\lambda} \frac{dx}{\beta(x)} \qquad (27)$$

For $\beta(x) = 3x^2/(16\pi^2) + O(x^3)$, whatever the bare contact coupling λ_0 at the asymptotically large Λ_s, at finite scales $\mu \sim M_h$ this gives $\lambda \sim 16\pi^2/(3L)$ with $L = \ln(\Lambda_s/M_h)$. It is this latter coupling which governs the residual interactions among the fluctuations with very small deviations from a purely quadratic potential for $\Lambda_s \to \infty$.

By introducing the W-mass $M_w = g_{\text{gauge}}\langle\Phi\rangle/2$ and with the notations of [30], a convenient way [29] to express these residual interactions in the scalar potential is ($r = M_h^2/4M_w^2 = K^2/g_{\text{gauge}}^2$)

$$U_{\text{scalar}} = \frac{1}{2}M_h^2 h^2 + \epsilon_1 r g_{\text{gauge}} M_w h(\chi^a\chi^a + h^2) + \frac{1}{8}\epsilon_2 r g_{\text{gauge}}^2(\chi^a\chi^a + h^2)^2 \qquad (28)$$

The two parameters ϵ_1 and ϵ_2, which are usually set to unity, take into account the basic difference $\lambda \neq 3K^2$, i.e.,

$$\epsilon_1^2 = \epsilon_2 = \frac{\lambda}{3K^2} \sim 1/L \tag{29}$$

Then, one can consider that corner of the parameter space [29], namely large K^2 but $M_h \ll \Lambda_s$, that does not exist in the conventional view where one assumes $\lambda = 3K^2$.

A possible objection to this scenario might concern its validity in the full gauge theory. In fact, the original calculation [31] in the unitary gauge could give the impression of the opposite view. Specifically, that with a heavy Higgs resonance of mass M_h, longitudinal $W_L W_L$ scattering is indeed governed by the large parameter $K^2 = M_h^2 / \langle \Phi \rangle^2$. Since this is an important point, we will repeat here the main argument of [29].

In the unitary-gauge calculation of $W_L W_L \to W_L W_L$ high-energy scattering, the lowest-order amplitude A_0 is formally $O(g_{\text{gauge}}^2)$ but one ends up with

$$A_0(W_L W_L \to W_L W_L) \sim \frac{3M_h^2 g_{\text{gauge}}^2}{4M_w^2} = \frac{3M_h^2}{\langle \Phi \rangle^2} = 3K^2 \tag{30}$$

In this chain, g_{gauge}^2 comes from the vertices. The $1/M_w^2$ originates from the external longitudinal polarizations $\epsilon_\mu^{(L)} \sim (k_\mu / M_w)$ and the factor M_h^2 emerges after expanding the Higgs field propagator

$$\frac{1}{s - M_h^2} \to \frac{1}{s}\left(1 + \frac{M_h^2}{s} + ...\right) \tag{31}$$

Then the leading $1/s$ contribution cancels against a similar term from the other diagrams (which otherwise would give an amplitude growing with s) and the M_h^2 from the expansion of the propagator is effectively "promoted" to the role of coupling constant. In this way, one gets exactly the same result as in a pure $\lambda \Phi^4$ theory with a contact coupling $\lambda_0 = 3K^2$.

However, this is only the tree approximation. To obtain the full result, let us observe that the Equivalence Theorem is a non perturbative statement which holds to all orders in the pure scalar self-interactions [28]. Therefore, we have not to worry to re-sum the infinite series of higher-order vector-boson graphs. However, from the $\chi\chi \to \chi\chi$ amplitude at a scale μ for $g_{\text{gauge}} = 0$

$$A(\chi\chi \to \chi\chi)\Big|_{g_{\text{gauge}}=0} \sim \lambda \sim \frac{1}{\ln(\Lambda_s / \mu)} \tag{32}$$

we can deduce the result for the longitudinal vector bosons in the $g_{\text{gauge}} \neq 0$ theory, i.e.,

$$A(W_L W_L \to W_L W_L) = [1 + O(g_{\text{gauge}}^2)]\, A(\chi\chi \to \chi\chi)\Big|_{g_{\text{gauge}}=0} \sim \lambda \sim \frac{1}{\ln(\Lambda_s / \mu)} \tag{33}$$

Then, in the present perspective of a large but finite Λ_s, where m_h and M_h now coexist and could be experimentally determined, at $\mu \sim M_h$ the putative strong interactions proportional to $\lambda_0 = 3K^2$ should actually be viewed as weak interactions controlled by the much smaller coupling

$$\lambda = \frac{3m_h^2}{\langle \Phi \rangle^2} = 3K^2 \frac{m_h^2}{M_h^2} \tag{34}$$

Analogously, the conventional very large width into longitudinal vector bosons computed with the coupling $\lambda_0 = 3K^2$, say $\Gamma^{\text{conv}}(M_h \to W_L W_L) \sim M_h^3 / \langle \Phi \rangle^2$, should instead be re-scaled by $\epsilon_1^2 = (\lambda / 3K^2) = m_h^2 / M_h^2$. This gives

$$\Gamma(M_h \to W_L W_L) \sim \frac{m_h^2}{M_h^2} \; \Gamma^{\text{conv}}(M_h \to W_L W_L) \sim M_h \frac{m_h^2}{\langle \Phi \rangle^2} \tag{35}$$

In this way, through the decays of the heavier state, the scalar coupling $\lambda = 3m_h^2/\langle \Phi \rangle^2 \sim 1/L$ could finally become visible.

6. Some Predictions for the LHC Experiments

Let us take seriously the idea of a BEH field with two vastly different mass scales, namely $m_h \sim 125$ GeV and $M_h \sim 700$ GeV. Is there any experimental signal from the LHC experiments? If so, what kind of phenomenology should we expect?

To address these questions, we will use a small but definite experimental evidence: the peak in the 4-lepton final state which is presently observed by the ATLAS Collaboration [32] for an invariant mass $\mu_{4l} = 700$ GeV. We emphasize that this should be taken seriously. In fact, an independent analysis of these data and their combination [33] with the corresponding ones of the CMS Collaboration indicates an evident excess, over the background, at the level of about 5 sigma.

Of course, the 4-lepton channel is only one decay channel of a hypothetical heavier BEH resonance and, for a more complete analysis, we should also consider the other final states. For instance the decay into two photons, a channel that in the past has been showing other intriguing evidence for the near energy $\mu_{\gamma\gamma} \sim 750$ GeV. However, the 4-lepton channel, has the advantage of being experimentally very clean and, just for this reason, is called the "golden" channel to detect a possible heavy BEH resonance. At the same time, as in ref. [34], the main effect can be analyzed at a very simple level. For this reason, one can meaningfully start from here.

Let us consider the peak in the number of events observed by ATLAS in the 4-lepton channel for an invariant mass $\mu_{4l} = 700$ GeV ($l = e, \mu$). From Figure 4a of [32] this corresponds to

$$3 \lesssim n^{\text{peak}}[4l] \lesssim 9 \qquad \text{ATLAS} - 700 \ \text{GeV} \tag{36}$$

above the very small background $n^{\text{bkg}} \sim 1$ event. By subtracting this background, we get

$$n^{\text{peak}}[4l] \sim 5 \pm 3 \quad (\text{non} - \text{bkg}) \ \text{EXP} \tag{37}$$

Since the ATLAS efficiency for reconstructed 4-lepton events at large transverse momentum is about 100%, for the given luminosity of 36.1 fb^{-1}, we obtain a peak cross-section

$$\sigma^{\text{peak}}(pp \to 4l) \sim (0.14 \pm 0.08) \ fb \tag{38}$$

For our estimates, we will assume the invariant mass $\mu_{4l} = 700$ GeV to be the same pole mass $M_h = 700$ GeV of our heavier excitation of the BEH field. Moreover, if we consider this as a relatively narrow resonance, the corrections due to its virtual propagation should be small [35] and one could approximate the result in terms of on-shell branching ratios as

$$\sigma(pp \to M_h \to 4l) \sim \sigma(pp \to M_h) \cdot B(M_h \to ZZ) \cdot 4B^2(Z \to l^+l^-) \tag{39}$$

In this relation, the $Z-$boson branching fraction into charged leptons is known precisely and one finds $4B^2(Z \to l^+l^-) \sim 0.0045$.

Concerning the other branching ratio $B(M_h \to ZZ)$, for $M_h = 700$ GeV, the only unconventional aspect of our picture concerns the coupling of the heavy BEH resonance to longitudinal vector bosons which is proportional to $\lambda = 3m_h^2/\langle \Phi \rangle^2 \sim 1/L$ and not to $3M_h^2/\langle \Phi \rangle^2$. Therefore, given a decay width $\Gamma(M_h \to ZZ)$, we could use the conventional estimate for $M_h = 700$ GeV [36,37]

$$\Gamma^{\text{conv}}(M_h \to ZZ) \sim 56.7 \ \text{GeV} \tag{40}$$

and, by replacing instead

$$\Gamma(M_h \to ZZ) \sim \frac{m_h^2}{M_h^2} \, \Gamma^{\text{conv}}(M_h \to ZZ) \tag{41}$$

obtain m_h as

$$m_h \sim \sqrt{\frac{\Gamma(M_h \to ZZ)}{56.7 \, \text{GeV}}} \, 700 \, \text{GeV} \tag{42}$$

Equivalently, given a value of m_h we can compute

$$\Gamma(M_h \to ZZ) \sim \frac{m_h^2}{(700 \, \text{GeV})^2} \, 56.7 \, \text{GeV} \tag{43}$$

Here, we will follow this latter strategy and assume $m_h = 125$ GeV which gives

$$\Gamma(M_h \to ZZ) \sim 1.8 \, \text{GeV} \tag{44}$$

Thus, to obtain $B(M_h \to ZZ)$, we only need to estimate the total decay width. Here, we will retain exactly the other contributions reported in the literature [36,37] for $M_h = 700$ GeV

$$\Gamma(M_h \to \text{fermions} + \text{gluons} + \text{photons}...) \sim 28 \, \text{GeV} \tag{45}$$

and the same dimensionless ratio

$$\frac{\Gamma(M_h \to WW)}{\Gamma(M_h \to ZZ)} \sim 2.03 \tag{46}$$

These input numbers (which have very small uncertainties) will then produce a total decay width

$$\Gamma(M_h \to all) \sim 28 \, \text{GeV} + 3.03 \, \Gamma(M_h \to ZZ) \sim 33.5 \, \text{GeV} \tag{47}$$

and a branching ratio

$$B(M_h \to ZZ) \sim \frac{1.8}{33.5} \sim 0.054 \tag{48}$$

Let us now consider the total cross-section $\sigma(pp \to M_h)$, for production of a heavy BEH resonance with mass $M_h \sim 700$ GeV. Here, the two main contributions derive from more elementary parton processes where two gluons or two vector bosons VV fuse to produce the heavy state M_h (here $VV = WW, ZZ$ would be emitted by two quarks inside the protons). For this reason, the two process are usually called Gluon-Gluon Fusion (GGF) and Vector-Boson Fusion (VBF) mechanisms, i.e.,

$$\sigma(pp \to M_h) \sim \sigma(pp \to M_h)_{\text{GGF}} + \sigma(pp \to M_h)_{\text{VBF}} \tag{49}$$

The traditional importance of the latter process for large M_h is understood by noticing that the $VV \to M_h$ process is the inverse of the $M_h \to VV$ decay and therefore $\sigma(pp \to M_h)_{\text{VBF}}$ can be expressed [38] as a convolution with the parton densities of the same BEH resonance decay width. Thus, once its coupling to longitudinal W's and Z's were proportional to $K^2 = M_h^2/\langle\Phi\rangle^2$, with a conventional width $\Gamma^{\text{conv}}(M_h \to WW + ZZ) \sim 172$ GeV for $M_h \sim 700$ GeV, the VBF mechanism would become important. However, this coupling is not present in our model, where instead we expect

$$\Gamma(M_h \to WW + ZZ) \sim \frac{m_h^2}{M_h^2} \, \Gamma^{\text{conv}}(M_h \to WW + ZZ) \sim 5.5 \, \text{GeV} \tag{50}$$

For this reason, the whole VBF will also be correspondingly reduced from its conventional value $\sigma^{\text{conv}}(pp \to M_h)_{\text{VBF}} = 250 \div 300 \, fb$, i.e.,

$$\sigma(pp \to M_h)_{\text{VBF}} \sim \frac{5.5}{172} \, \sigma^{\text{conv}}(pp \to M_h)_{\text{VBF}} \lesssim 10 \, fb \tag{51}$$

This is much smaller than the uncertainty in the pure GGF contribution and will be ignored in the following.

In the end, the GGF term. Here, we will separately adopt two slightly different estimates. On the one hand, the value $\sigma(pp \rightarrow M_h)_{\text{GGF}} = 800(80)$ fb of ref. [36] and on the other hand, the value $\sigma(pp \rightarrow M_h)_{\text{GGF}} = 1078(150)$ fb of ref. [37]. These values refer to $\sqrt{s} = 14$ TeV and will be re-scaled by about -12% for the present center of mass energy $\sqrt{s} = 13$ TeV. In the two cases, the errors take into account uncertainties in the normalization scale and in the parametrization of the parton distributions.

Altogether, for $B(M_h \rightarrow ZZ) = 0.054$ and $4B^2(Z \rightarrow l^+l^-) \sim 0.0045$, our predictions for the 4-lepton cross-section and the number of events (for luminosity of 36.1 fb^{-1} and 139 fb^{-1}) are reported in Table 1.

Table 1. For $M_h = 700$ GeV and $m_h = 125$ GeV, we report our predictions for the peak cross-section $\sigma(pp \rightarrow 4l)$ and the number of events at two values of the luminosity. The two total cross sections are our extrapolation to $\sqrt{s} = 13$ TeV of the values in [36,37] for $\sqrt{s} = 14$ TeV. As explained in the text, only the GGF mechanism is relevant in our model.

$\sigma(pp \rightarrow M_h)$	$\sigma(pp \rightarrow 4l)$	$n[4l](\mathcal{L} = 36.1\,fb^{-1})$	$n[4l](\mathcal{L} = 139\,fb^{-1})$
700(70) fb	0.17(2) fb	6.1 ± 0.6	23.6 ± 2.4
950(150) fb	0.23(4) fb	8.3 ± 1.3	32.1 ± 5.1

From this comparison we deduce that without introducing any free parameter, our model can easily reproduce the presently observed number of events $n[4l] \sim 5 \pm 3$. This is why, our hypothetical new resonance could naturally fit with the ATLAS peak. At present, this is the only possible conclusion and a real test of our picture is postponed to the analysis of the entire statistics $\mathcal{L} = 139\ fb^{-1}$. If the new $M_h \sim 700$ GeV were really there, the peak should become four times higher but remain well above the background which is very small at that energy. Thus, the profile of the resonance should become visible and direct determinations of the total decay width should be feasible. An experimental result $\Gamma^{\text{exp}}(M_h \rightarrow all) = 33 \div 34$ GeV would favor an experimental branching ratio $B^{\text{exp}}(M_h \rightarrow ZZ)$ close to our reference value 0.054 and, therefore, improve the agreement of our smaller m_h with the value 125 GeV which is measured directly at LHC. Thus, the description of SSB given here would find a first experimental confirmation.

Author Contributions: Conceptualization, M.C. and L.C.; methodology, M.C. and L.C.; software, M.C. and L.C.; validation, M.C. and L.C.; formal analysis, M.C. and L.C.; investigation, M.C. and L.C.; resources, M.C. and L.C.; data curation, M.C. and L.C.; writing—original draft preparation, M.C. and L.C.; writing—review and editing, M.C. and L.C.; visualization, M.C. and L.C.; supervision, M.C. and L.C. All authors have read and agreed to the published version of the manuscript.

Funding: This research received no external funding.

Acknowledgments: This paper is dedicated to Yves Brihaye in the occasion of his 65th birthday.

Conflicts of Interest: The authors declare no conflict of interest.

References

1. Englert, F.; Brout, R. Broken Symmetry and the Mass of Gauge Vector Mesons. *Phys. Rev. Lett.* **1964**, *13*, 321–323. [CrossRef]
2. Higgs, P.W. Broken symmetries, massless particles and gauge fields. *Phys. Lett.* **1964**, *12*, 132–133. [CrossRef]
3. Aad, G.; Abajyan, T.; Abbott, B.; Abdallah, J.; Khalek, S.A.; Abdelalim, A.A.; Aben, R.; Abi, B.; Abolins, M.; AbouZeid, O.S.; et al. Observation of a new particle in the search for the Standard Model Higgs boson with the ATLAS detector at the LHC. *Phys. Lett. B* **2012**, *716*, 1–29. [CrossRef]
4. Chatrchyan, S.; Khachatryan, V.; Sirunyan, A.M.; Tumasyan, A.; Adam, W.; Aguilo, E.; Bergauer, T.; Dragicevic, M.; Erö, J.; Fabjan, C.; et al. Observation of a New Boson at a Mass of 125 GeV with the CMS Experiment at the LHC. *Phys. Lett. B* **2012**, *716*, 30–61. [CrossRef]

5. Coleman, S.R.; Weinberg, E.J. Radiative Corrections as the Origin of Spontaneous Symmetry Breaking. *Phys. Rev. D* **1973**, *7*, 1888–1910. [CrossRef]

6. Consoli, M.; Stevenson, P.M. Physical mechanisms generating spontaneous symmetry breaking and a hierarchy of scales. *Int. J. Mod. Phys. A* **2000**, *15*, 133. [CrossRef]

7. Barnes, T.; Ghandour, G.I. Variational Treatment of the Effective Potential and Renormalization in Fermi-Bose Interacting Field Theories. *Phys. Rev. D* **1980**, *22*, 924. [CrossRef]

8. Stevenson, P.M. The Gaussian Effective Potential. 2. Lambda phi**4 Field Theory. *Phys. Rev. D* **1985**, *32*, 1389–1408. [CrossRef]

9. Consoli, M.; Ciancitto, A. Indications of the occurrence of spontaneous symmetry breaking in massless $\lambda\phi^4$. *Nucl. Phys. B* **1985**, *254*, 653–677. [CrossRef]

10. Brihaye, Y.; Consoli, M. Spontaneous symmetry breaking in an O(2) invariant scalar theory. *Phys. Lett. B* **1985**, *157*, 48–52. [CrossRef]

11. Brihaye, Y.; Consoli, M. Gaussian Quantization, O(N) Theory and the Goldstone Theorem. *Nuovo Cim. A* **1986**, *94*, 1–14. [CrossRef]

12. Stancu, I.; Stevenson, P.M. Second Order Corrections to the Gaussian Effective Potential of $\lambda\phi^4$ Theory. *Phys. Rev. D* **1990**, *42*, 2710–2725. [CrossRef] [PubMed]

13. Cea, P.; Tedesco, L. Perturbation theory with a variational basis: The Generalized Gaussian effective potential. *Phys. Rev. D* **1997**, *55*, 4967–4989. [CrossRef]

14. Consoli, M.; Stevenson, P.M. The Nontrivial effective potential of the 'trivial' $\lambda\phi^4$ theory: A Lattice test. *Z. Phys. C* **1994**, *63*, 427–436. [CrossRef]

15. Consoli, M.; Stevenson, P.M. Mode-dependent field renormalization and triviality in $\lambda\phi^4$ theory. *Phys. Lett. B* **1997**, *391*, 144–149. [CrossRef]

16. Feinberg, G.; Sucher, J. Long-Range Forces from Neutrino-Pair Exchange. *Phys. Rev.* **1968**, *166*, 1638–1644. [CrossRef]

17. Feinberg, G.; Sucher, J.; Au, C.K. The Dispersion Theory of Dispersion Forces. *Phys. Rept.* **1989**, *180*, 83. [CrossRef]

18. Stevenson, P.M. The Long-range interaction in massless $\lambda\phi^4$ theory. *Mod. Phys. Lett. A* **2009**, *24*, 261–271. [CrossRef]

19. Lundow, P.H.; Markström, K. Critical behavior of the Ising model on the four-dimensional cubic lattice. *Phys. Rev. E* **2009**, *80*, 031104. [CrossRef]

20. Lundow, P.H.; Markstrom, K. Non-vanishing boundary effects and quasi-first order phase transitions in high dimensional Ising models. *Nucl. Phys. B* **2011**, *845*, 120–139. [CrossRef]

21. Akiyama, S.; Kuramashi, Y.; Yamashita, T.; Yoshimura, Y. Phase transition of four-dimensional Ising model with higher-order tensor renormalization group. *Phys. Rev. D* **2019**, *100*, 054510. [CrossRef]

22. Consoli, M.; Cosmai, L. The mass scales of the Higgs field. *Int. J. Mod. Phys. A* **2020**, *35*, 2050103. [CrossRef]

23. Chanowitz, M.S. Z-prime Bosons, the NuTeV Anomaly, and the Higgs Boson Mass. *Phys. Atom. Nucl.* **2010**, *73*, 680–688. [CrossRef]

24. Cea, P.; Consoli, M.; Cosmai, L. Indications on the Higgs boson mass from lattice simulations. *Nucl. Phys. Proc. Suppl.* **2004**, *129*, 780–782. [CrossRef]

25. Cea, P.; Cosmai, L. The Higgs boson: From the lattice to LHC. *ISRN High Energy Phys.* **2012**, *2012*, 637950. [CrossRef]

26. Cornwall, J.M.; Levin, D.N.; Tiktopoulos, G. Derivation of Gauge Invariance from High-Energy Unitarity Bounds on the s Matrix. *Phys. Rev. D* **1974**, *10*, 1145. [CrossRef]

27. Chanowitz, M.S.; Gaillard, M.K. The TeV Physics of Strongly Interacting W's and Z's. *Nucl. Phys. B* **1985**, *261*, 379–431. [CrossRef]

28. Bagger, J.; Schmidt, C. Equivalence Theorem Redux. *Phys. Rev. D* **1990**, *41*, 264. [CrossRef]

29. Castorina, P.; Consoli, M.; Zappala, D. An Alternative heavy Higgs mass limit. *J. Phys. G* **2008**, *35*, 075010. [CrossRef]

30. Veltman, M.J.G.; Yndurain, F.J. Radiative corrections to W W scattering. *Nucl. Phys. B* **1989**, *325*, 1–17. [CrossRef]

31. Lee, B.W.; Quigg, C.; Thacker, H.B. Weak Interactions at Very High-Energies: The Role of the Higgs Boson Mass. *Phys. Rev. D* **1977**, *16*, 1519. [CrossRef]

32. Aaboud, M.; Aad, G.; Abbott, B.; Abdinov, O.; Abeloos, B.; Abidi, S.H.; AbouZeid, O.S.; Abraham, N.L.; Abramowicz, H.; Abreu, H.; et al. Search for heavy ZZ resonances in the $\ell^+\ell^-\ell^+\ell^-$ and $\ell^+\ell^-\nu\bar{\nu}$ final states using proton–proton collisions at $\sqrt{s} = 13$ TeV with the ATLAS detector. *Eur. Phys. J. C* **2018**, *78*, 293. [CrossRef] [PubMed]

33. Cea, P. Evidence of the true Higgs boson H_T at the LHC Run 2. *Mod. Phys. Lett. A* **2019**, *34*, 1950137. [CrossRef]

34. Consoli, M.; Cosmai, L. A resonance of the Higgs field at 700 GeV and a new phenomenology. *arXiv* **2020**, arXiv:2007.10837. Available online: https://arxiv.org/abs/2007.10837 (accessed on 20 July 2020).

35. Goria, S.; Passarino, G.; Rosco, D. The Higgs Boson Lineshape. *Nucl. Phys. B* **2012**, *864*, 530–579. [CrossRef]

36. Djouadi, A. The Anatomy of electro-weak symmetry breaking. I: The Higgs boson in the standard model. *Phys. Rept.* **2008**, *457*, 1–216. [CrossRef]

37. Dittmaier, S.; Mariotti, C.; Passarino, G.; Tanaka, R.; Baglio, J.; Bolzoni, P.; Boughezal, R.; Brein, O.; Collins-Tooth, C.; Dawson, S.; et al. Handbook of LHC Higgs Cross Sections: 1. Inclusive Observables. *arXiv* **2011**, arXiv:1101.0593. [CrossRef]

38. Kane, G.L.; Repko, W.; Rolnick, W. The Effective W+-, Z0 Approximation for High-Energy Collisions. *Phys. Lett. B* **1984**, *148*, 367–372. [CrossRef]

Article

Asymptotically Flat, Spherical, Self-Interacting Scalar, Dirac and Proca Stars

Carlos A. R. Herdeiro and Eugen Radu *

Departamento de Matemática da Universidade de Aveiro and Centre for Research and Development in Mathematics and Applications (CIDMA), Campus de Santiago, 3810-183 Aveiro, Portugal; herdeiro@ua.pt
* Correspondence: eugen.radu@ua.pt

Received: 14 November 2020; Accepted: 1 December 2020; Published: 8 December 2020

Abstract: We present a comparative analysis of the self-gravitating solitons that arise in the Einstein–Klein–Gordon, Einstein–Dirac, and Einstein–Proca models, for the particular case of static, spherically symmetric spacetimes. Differently from the previous study by Herdeiro, Pombo and Radu in 2017, the matter fields possess suitable self-interacting terms in the Lagrangians, which allow for the existence of Q-ball-type solutions for these models in the flat spacetime limit. In spite of this important difference, our analysis shows that the high degree of universality that was observed by Herdeiro, Pombo and Radu remains, and various spin-independent common patterns are observed.

Keywords: solitons; boson stars; Dirac stars

1. Introduction and Motivation

1.1. General Remarks

The (modern) idea of solitons, as extended particle-like configurations, can be traced back (at least) to Lord Kelvin, around one and half centuries ago, who proposed that atoms are made of vortex knots [1]. However, the first explicit example of solitons in a relativistic field theory was found by Skyrme [2,3], (almost) one hundred years later. The latter, dubbed *Skyrmions*, exist in a model with four (real) scalars that are subject to a constraint. They were proposed as field theory realizations of baryons; nonetheless, Skyrmions capture the basic properties of a generic soliton.

In the context of this work, solitons are defined as spatially localized, singularity-free solutions of a field theory model, which possess a finite mass and angular momentum (the fields may possess a non-trivial time-dependence, as for the solitons in this work. However, this dependence is absent at the level of the energy-momentum tensor. Additionally, in our definition, we shall ignore the issue of stability and do not impose the solutions possessing a topological charge). As already suspected in the literature before Skyrme's seminal work (in a historical setting, a concrete realization of the idea of (relativistic) solitons was put forward in 1939 by Rosen [4], who looked for localized configurations which could be used as representations of extended particles), many non-linear field theories possess such solutions. The study of solitons has attracted a lot of attention in the last 50 years; they have important applications in various contexts, ranging from, e.g., the models of condensed matter physics to high energy physics and cosmology. Moreover, solitons are also significant for a nonperturbative quantum description of states. Famous examples of solitons include the sphalerons in the Standard Model [5] and the magnetic monopoles [6,7] in various extensions thereof.

A review of soliton physics in the non-gravitating case can be found in the textbooks [8,9]; also see the reviews [10,11]. In this work, we shall focus on a special class of solitons in four spacetime dimensions that are time-dependent (at the level of the matter field), but non-dissipative (with a time independent spacetime geometry), and carrying a Noether charge Q that is associated with a global $U(1)$ symmetry. We shall restrict ourselves to models with a single matter field (except for the fermionic case) and possessing a standard kinetic term. These solitons do not possess topological properties, but the field theory Lagrangian must have suitable self-interactions.

Three different models are considered here: (ungauged) field theories describing particles of spin $s = 0, 1/2, 1$, with the goal of providing a comparative analysis of their spherically symmetric solitonic solutions. As such, this work is an extension of that in Ref. [12], which dealt with the same matter fields, but without self-interactions. The latter introduce a novel feature: self-interactions allow for the existence of solitonic solutions, even on a fixed flat spacetime background, i.e., ignoring gravity. As we shall see, self-interactions lead to a more intricate landscape of (gravitating) solutions, with some new qualitative features when compared to the picture that is revealed in [12].

Historically, in the non-gravitating case, solitons in a complex scalar field model have been known for more than 50 years, being dubbed $Q - balls$ by Coleman [13] (also see the previous work [14]). These are central for our discussion, sharing most of the basic features with the higher spin generalizations. Within a Dirac field model with a positive quartic self-interaction term, Ivanenko [15], Weyl [16], Heisenberg [17], and Finkelstein et al. [18,19] have made early attempts to construct particle-like solutions. However, the first rigorous numerical study of such configurations was been done by Soler in 1976 [20]. The case of a (Abelian) vector field has been ignored until recently; the study of Q-ball-like solutions was initiated by Loginov [21].

In all three cases, a large literature has grown based on these early studies, which includes generalizations in various directions. A particularly interesting case concerns the study of backreaction of these solutions on the spacetime geometry. This is a legitimate question, since the existence of solitons relies on the nonlinearity of the field theory and General Relativity (GR) is intrinsically highly nonlinear. Therefore, gravitational interactions could significantly alter the flat spacetime solitons.

The results in the literature prove that all flat space solitons survive when including gravity effects; however, a more complicated picture emerges when compared to the non-self-interacting case. The study of gravitating generalization of the flat spacetime Q-balls has started with Friedberg et al. [22]. The higher spin fields are less studied; GR generalization of the self-interacting spinors were only considered recently by Dzhunushaliev and Folomeev [23], while the gravitating Q-Proca stars have been studied by Brihaye et al. [24] and Minamitsuji [25].

The main purpose of this work is to review various existing results on this type of solitons, putting them together under a consistent set of notations and conventions. A comparative study is presented, which starts with the generic framework and scaling symmetries (Section 2). There and afterwards, the mathematical description of each of the three models is made in parallel in order to emphasise their similarities. The basic properties of solutions in a flat spacetime background are discussed in Section 3, where we adapt Deser's argument [26] to find virial identities that yield some insight on the existence of such configurations. Gravity is included in Section 4; a curved spacetime generalization of the Deser-type virial identities is also discussed. Extending the models' framework to include GR effects allows for the possibility to add a black hole (BH) horizon at the center of soliton, a possibility that is indeed realized for other well-known solutions, such as Skyrmions, magnetic monopoles, and sphalerons (see e.g., the review work [27]). However, as we shall discuss in Section 4 (and Appendix B), the situation is different for the considered (spherically symmetric) solutions and there are no BH generalizations, regardless of the presence of self-interactions. We also discuss the physical interpretation of the fermionic solutions. Section 5 presents concluding remarks and some open questions.

1.2. Conventions

Throughout, we set $c = \hbar = 1$ and adopt a metric signature $+2$. Greek letters $\alpha, \beta, \gamma \ldots$ are used for coordinate indices, whereas latin letters $a, b, c, \ldots$ are used for tetrad basis indices. Symmetrization and antisymmetrization of indices is denoted with round and square brackets, $()$ and $[]$, respectively. We use ∂_μ, ∇_μ, and $\hat{D}_\mu$ to denote partial, covariant, and spinor derivatives, respectively.

The flat spacetime metric in spherical coordinates reads

$$ds^2 = -dt^2 + dr^2 + r^2(d\theta^2 + \sin^2\theta d\varphi^2), \tag{1}$$

with $x^0 = t$, $x^1 = r$, $x^2 = \theta$, and $x^3 = \varphi$.

The conventions for scalars are those presented in Ref. [28]. In the Proca field case, we shall use the notation and conventions in [29,30]. For fermions, we shall follow the framework (including the definitions and conventions) in [31], as reviewed in Appendix A.

For the matter content, we shall consider three different fields ψ, with the specific form:

- **spin 0:**
  Φ is a complex scalar field, which is equivalent to a model with two real scalar fields, Φ^R, Φ^I, via $\Phi = \Phi^R + i\Phi^I$.
- **spin 1/2:**
  $\Psi^{(A)}$ are massive spinors, with four complex components, the index (A) corresponding to the number of copies of the Lagrangian. For a spherically symmetric configuration, one should consider (at least) two spinors ($A = 1,2$), with equal mass μ. A model with a single spinor necessarily possesses a nonzero angular momentum density and it cannot be spherically symmetric.
- **spin 1:**
  $\mathcal{A}$ is a complex 4-potential, with the field strength $\mathcal{F} = d\mathcal{A}$. Again, the model can be described in terms of two real vector fields, $\mathcal{A} = \mathcal{A}^R + i\mathcal{A}^I$.

Some relations get a simpler form by defining

$$\psi^2 \equiv \{\Phi^*\Phi; \ \mathcal{A}_\alpha \bar{\mathcal{A}}^\alpha; \ i\overline{\Psi}^{(A)}\Psi^{(A)}\}, \tag{2}$$

for spin $0, 1/2$,and 1, respectively, while we denote $\psi^4 \equiv (\psi^2)^2$ and $\psi^6 \equiv (\psi^2)^3$.

The numerical construction of the solutions reported in this work is standard. In our approach, we use a Runge–Kutta ordinary differential equation solver. For each model, we evaluate the initial conditions that are close to the origin at $\epsilon = 10^{-6}$ for global tolerance 10^{-14}, adjusting for shooting parameters (which are some constants that enter the near origin expansion of the solutions) and integrating towards $r \to \infty$. The accuracy of the solutions was also monitored by computing virial relations that were satisfied by these systems, as discussed below. For a given set of input parameters, the solutions form a discrete set labelled by the number of nodes, n, of (some of) the matter function(s). Only fundamental states, which have the minimal number of radial nodes, are considered in this work.

2. The General Framework

2.1. The Action and Field Equations

We consider Einstein's gravity that is minimally coupled to spin$-s$ matter fields $\psi_{(s)} = \{\Phi; \Psi; \mathcal{A}\}$, with $s = 0, 1/2, 1$. The action reads

$$S = \frac{1}{4\pi}\int d^4x \sqrt{-g}\left[\frac{1}{4G}R - L_{(s)}\right], \tag{3}$$

where

$$L_{(s)} = L_{(s)}^{(0)} + U_{(s)}^{(\text{int})} ; \tag{4}$$

the kinetic part of the matter Lagrangians reads, respectively,

$$\text{scalar}: \quad L^{(0)}_{(0)} = g^{\alpha\beta}\Phi^{*}_{,\alpha}\Phi_{,\beta},$$

$$\text{Dirac}: \quad L^{(A)(0)}_{(1/2)} = i\left[\frac{1}{2}\left(\{\hat{D}\overline{\Psi}^{(A)}\}\Psi^{(A)} - \overline{\Psi}^{(A)}\hat{D}\Psi^{(A)}\right)\right], \tag{5}$$

$$\text{Proca}: \quad L^{(0)}_{(1)} = \frac{1}{4}\mathcal{F}_{\alpha\beta}\bar{\mathcal{F}}^{\alpha\beta}.$$

$U^{(\text{int})}_{(s)}$ in (4) is a potential term, which includes self-interactions. In what follows, we shall consider the simplest form of $U^{(\text{int})}_{(s)}$, which allows for solitonic solutions in a flat spacetime background, with polynomial terms. As such, apart from the mass term, $U^{(\text{int})}_{(s)}$ contains quartic and sextic terms only. Afterwards, all three cases can be treated in a unitary way by defining

$$U^{(\text{int})}_{(s)} = \mathcal{M}\psi^{2} - \lambda\psi^{4} + \nu\psi^{6} \equiv U, \tag{6}$$

where we have denoted

$$\mathcal{M} \equiv \left\{\mu^{2};\ \mu;\ \frac{1}{2}\mu^{2}\right\}, \tag{7}$$

for spin $0, 1/2$ and 1, respectively. In all cases, μ corresponds to the mass of the elementary quanta of the field(s). Additionally, $\lambda,\ \nu$ are the input parameters of the theory, which are not fixed *a priori*.

We shall also denote

$$\dot{U} \equiv \frac{\partial U}{\partial\psi^{2}} = \mathcal{M} - 2\lambda\psi^{2} + 3\nu\psi^{4}, \qquad \ddot{U} \equiv \frac{\partial\dot{U}}{\partial\psi^{2}} = -2\lambda + 6\nu\psi^{2}. \tag{8}$$

Extremizing the action (3) leads to the Einstein field equations

$$R_{\alpha\beta} - \frac{1}{2}Rg_{\alpha\beta} = 2G\,T_{\alpha\beta}, \tag{9}$$

where the energy–momentum tensor reads, respectively,

$$\text{scalar}: \quad T_{\alpha\beta} = \Phi^{*}_{,\alpha}\Phi_{,\beta} + \Phi^{*}_{,\beta}\Phi_{,\alpha} - g_{\alpha\beta}L_{(0)},$$

$$\text{Dirac}: T_{\alpha\beta} = \sum_{A} T^{(A)}_{\alpha\beta}, \ \text{with}\ T^{(A)}_{\alpha\beta} = -\frac{i}{2}\left[\overline{\Psi}^{(A)}\gamma_{(\alpha}\hat{D}_{\beta)}\Psi^{(A)} - \left\{\hat{D}_{(\alpha}\overline{\Psi}^{(A)}\right\}\gamma_{\beta)}\Psi^{(A)} - g_{\alpha\beta}L_{(1/2)}\right],$$

$$\text{Proca}: T_{\alpha\beta} = \frac{1}{2}(\mathcal{F}_{\alpha\sigma}\bar{\mathcal{F}}_{\beta\gamma} + \bar{\mathcal{F}}_{\alpha\sigma}\mathcal{F}_{\beta\gamma})g^{\sigma\gamma} + \dot{U}(\mathcal{A}_{\alpha}\bar{\mathcal{A}}_{\beta} + \bar{\mathcal{A}}_{\alpha}\mathcal{A}_{\beta}) - g_{\alpha\beta}L_{(1)}, \tag{10}$$

and to the matter field equations,

$$\text{scalar}: \quad (\nabla^{2} - \dot{U})\Phi = 0,$$

$$\text{Dirac}: \quad (\hat{D} - \dot{U})\Psi^{(A)} = 0, \tag{11}$$

$$\text{Proca}: \quad \frac{1}{2}\nabla_{\alpha}\mathcal{F}^{\alpha\beta} - \dot{U}\mathcal{A}^{\beta} = 0.$$

In the Proca case, taking the four-divergence of the field Equation (11) results in a generalization of the Lorenz condition, which is a dynamical requirement, rather than a gauge choice. This condition takes a particularly simple form in the Ricci-flat case, with

$$(\nabla_{\alpha}\mathcal{A}^{\alpha})\dot{U} + \ddot{U}\nabla_{\alpha}(\bar{\mathcal{A}}_{\beta}\mathcal{A}^{\beta}) = 0. \tag{12}$$

In all cases, the action of the matter fields ψ possesses a global $U(1)$ invariance, under the transformation $\psi \to e^{i\alpha}\psi$, with α constant. This implies the existence of a conserved four-current, which reads

$$
\begin{aligned}
\text{scalar}: \ & j^\alpha = -i(\Phi^* \partial^\alpha \Phi - \Phi \partial^\alpha \Phi^*), \\
\text{Dirac}: \ & j^\alpha = \bar{\Psi}\gamma^\alpha \Psi, \\
\text{Proca}: \ & j^\alpha = \frac{i}{2}\left[\bar{\mathcal{F}}^{\alpha\beta}\mathcal{A}_\beta - \mathcal{F}^{\alpha\beta}\bar{\mathcal{A}}_\beta\right].
\end{aligned}
\tag{13}
$$

This current is conserved via the field equations,

$$
j^\alpha{}_{;\alpha} = 0.
\tag{14}
$$

It follows that integrating the timelike component of this four-current in a spacelike slice Σ yields a conserved quantity—the Noether charge:

$$
Q = \int_\Sigma j^t .
\tag{15}
$$

2.2. The Ansätze and Explicit Equations

2.2.1. The Metric and Matter Fields

The spherically symmetric configurations are most conveniently studied in Schwarzschild-like coordinates, within the following metric Ansatz:

$$
ds^2 = -N(r)\sigma^2(r)dt^2 + \frac{dr^2}{N(r)} + r^2(d\theta^2 + \sin^2\theta\, d\varphi^2), \quad \text{with } N(r) \equiv 1 - \frac{2m(r)}{r}.
\tag{16}
$$

The matter field Ansatz, which is compatible with a spherically symmetric geometry, reads:

$$
\text{scalar}: \ \Phi = \phi(r)e^{-iwt},
\tag{17}
$$

introducing a single real function $\phi(r)$;

$$
\text{Proca}: \ \mathcal{A} = \left[F(r)dt + iH(r)dr\right]e^{-iwt},
\tag{18}
$$

introducing two real potentials $F(r)$ and $H(r)$.

The case of a Dirac field is more involved. For a spherically symmetric configurations, we have to consider two Dirac fields, $A = 1, 2$, with

$$
\text{Dirac}: \ \Psi^{(A)} = e^{-iwt}\mathcal{R}^{(A)}(r) \otimes \Theta^{(A)}(\theta, \varphi),
\tag{19}
$$

where

$$
\mathcal{R}^{(1)} = -i\mathcal{R}^{(2)} = \begin{bmatrix} z(r) \\ -i\bar{z}(r) \end{bmatrix}, \quad
\Theta^{(1)} = \begin{bmatrix} -\kappa\sin\frac{\theta}{2} \\ \cos\frac{\theta}{2} \end{bmatrix}e^{i\frac{\varphi}{2}}, \quad
\Theta^{(2)} = \begin{bmatrix} \kappa\cos\frac{\theta}{2} \\ \sin\frac{\theta}{2} \end{bmatrix}e^{-i\frac{\varphi}{2}},
$$

with $\kappa = \pm 1$ and $z(r)$ a complex function. In what follows, we shall only consider the case $\kappa = 1$ (note that qualitatively similar solutions with $\kappa = -1$ also exist). Additionally, in the above Ansatz, it is convenient to define

$$
z(r) = e^{i\pi/4}f(r) - e^{-i\pi/4}g(r),
\tag{20}
$$

where $f(r)$ and $g(r)$ are two real functions, a choice that simplifies the equations.

The reason that choose two independent $s = 1/2$ fields, with the Ansatz (19) and (20), is the following. For both spinors, the *individual* energy-momentum tensors are not spherically symmetric (this feature is present regardless of the self-interaction potential in the Lagrangian), since $T_\varphi^{t(A)} \neq 0$ (while the other nonzero components $T_r^{r(A)}$, $T_\theta^{\theta(A)} = T_\varphi^{\varphi(A)}$ and $T_t^{t(A)}$ only depend on r). However, $T_\varphi^{t(1)} + T_\varphi^{t(2)} = 0$, such that the full configuration is spherically symmetric, being compatible with the line element (16).

Within this framework, the explicit expressions of ψ^2, as defined in (2), are

$$\psi^2 = \left\{ \phi^2, \ 4(g^2 - f^2), \ H^2 N - \frac{F^2}{N\sigma^2} \right\} \tag{21}$$

for spin $0, 1/2$ and 1, respectively. In all cases, $w > 0$ is the frequency of the matter field.

2.2.2. The Explicit Equations

The equations for the mass function $m(r)$ read, respectively,

$$\text{scalar}: \ m' = Gr^2 \left(N\phi'^2 + \frac{w^2\phi^2}{N\sigma^2} + U \right),$$

$$\text{Dirac}: \ m' = 2Gr^2 \left(4\sqrt{N}(gf' - fg') + \frac{8fg}{r} + U \right), \tag{22}$$

$$\text{Proca}: \ m' = Gr^2 \left[\frac{(F' - wH)^2}{2\sigma^2} + (\mu^2 - 6\lambda\mathcal{A}^2 + 10\nu\mathcal{A}^4)\frac{F^2}{2N\sigma^2} + \frac{U}{\mathcal{A}^2} NG^2 \right].$$

The equations for the metric function $\sigma(r)$ read, respectively,

$$\text{scalar}: \ \frac{\sigma'}{\sigma} = 2Gr \left(\phi'^2 + \frac{w^2\phi^2}{N^2\sigma^2} \right),$$

$$\text{Dirac}: \ \frac{\sigma'}{\sigma} = 8G\frac{r}{\sqrt{N}} \left(gf' - fg' + \frac{w(f^2 + g^2)}{N\sigma} \right), \tag{23}$$

$$\text{Proca}: \ \frac{\sigma'}{\sigma} = \frac{2Gr}{N} \left(H^2N + \frac{F^2}{N\sigma^2} \right) \ddot{U}.$$

The equations for the matter fields are

$$\text{scalar}: \ \phi'' + \left(\frac{2}{r} + \frac{N'}{N} + \frac{\sigma'}{\sigma} \right) \phi' + \frac{w^2}{N^2\sigma^2}\phi - \ddot{U}\frac{\phi}{N} = 0,$$

$$\text{Dirac} - f: \ f' + \left(\frac{N'}{4N} + \frac{\sigma'}{2\sigma} + \frac{1}{r\sqrt{N}} + \frac{1}{r} \right) f - \frac{wg}{N\sigma} + \frac{g}{\sqrt{N}}\ddot{U} = 0,$$

$$\text{Dirac} - g: \ g' + \left(\frac{N'}{4N} + \frac{\sigma'}{2\sigma} - \frac{1}{r\sqrt{N}} + \frac{1}{r} \right) f + \frac{wf}{N\sigma} + \frac{f}{\sqrt{N}}\ddot{U} = 0, \tag{24}$$

$$\text{Proca} - F: \ \ F' - wH + \frac{2N\sigma^2 H}{w}\ddot{U} = 0,$$

$$\text{Proca} - H: \ \ \frac{d}{dr}\left\{ \frac{r^2[wH - F']}{\sigma} \right\} + \frac{2r^2F}{N\sigma}\ddot{U} = 0.$$

An additional supplementary constraint is also present, which is a second order equation for the metric functions $m(r)$ and $\sigma(r)$, with first order derivatives of matter fields. However, this equation is a differential consequence of the above field equations; it is used to check the accuracy of the numerical results.

Let us remark that the above equations can be derived from the following effective action

$$S_{\text{eff}} = \int dr \mathcal{L}_{\text{eff}}, \qquad \text{with } \mathcal{L}_{\text{eff}} = \frac{1}{G}\sigma m' - \mathcal{L}_{(s)}, \tag{25}$$

where

$$\mathcal{L}_{(0)} = r^2\sigma\left(N\phi'^2 - \frac{w^2\phi^2}{N\sigma^2} + U\right),$$

$$\mathcal{L}_{(1/2)} = 8r^2\sigma\left(\sqrt{N}(gf' - fg') - \frac{w}{\sigma\sqrt{N}}(f^2 + g^2) + \frac{2fg}{r} + \frac{1}{4}U\right), \tag{26}$$

$$\mathcal{L}_{(1)} = r^2\sigma\left(-\frac{(F' - wH)^2}{2\sigma^2} + U\right).$$

Given the above framework, the energy density measured by a static observer, $\rho = -T^t_t$, is

$$\text{scalar}: \rho = N\phi'^2 + \frac{w^2}{N\sigma^2}\phi^2 + U,$$

$$\text{Dirac}: \rho = 8\left((gf' - fg')\sqrt{N} + \frac{2fg}{r} + \frac{1}{4}U\right), \tag{27}$$

$$\text{Proca}: \rho = \frac{(F' - wH)^2}{2\sigma^2} + \frac{U}{\mathcal{A}^2}NH^2 + \frac{F^2}{N\sigma^2}\left(\frac{1}{2}\mu^2 - 3\lambda\mathcal{A}^2 + 5\nu\mathcal{A}^4\right).$$

The mass of the flat space solutions is computed as the integral of ρ,

$$M = \int dr r^2 \rho, \tag{28}$$

while, in the self-gravitating case, it can be read off from the asymptotic behaviour of the g_{tt} metric potential

$$g_{tt} = -N\sigma^2 = -1 + \frac{2MG}{r} + \dots \tag{29}$$

As for the Noether charge, it reads

$$\text{scalar}: Q = 2w\int_0^\infty dr\, r^2 \frac{\phi^2}{N\sigma},$$

$$\text{Dirac}: Q = 4\int_0^\infty dr\, r^2 \frac{(f^2 + g^2)}{\sqrt{N}}, \tag{30}$$

$$\text{Proca}: Q = 2\int_0^\infty dr\, r^2 \frac{(wH - F')N}{\sigma}.$$

2.3. Units and Scaling Symmetries

The matter fields in action (3) have the following dimensions: (with $L = lenght$):

$$\text{scalar}: [\Phi] = \frac{1}{L},$$

$$\text{Dirac}: [\Psi] = \frac{1}{L^{3/2}}, \tag{31}$$

$$\text{Proca}: [\mathcal{A}_a] = \frac{1}{L}.$$

Additionally, the coupling constants that enter the potential are (generically) dimension-full, with:

$$\text{scalar}: \ [\mu] = \frac{1}{L}, \ [\lambda] = L^0, \ [v] = L^2,$$

$$\text{Dirac}: \ [\mu] = \frac{1}{L}, \ [\lambda] = L^2, \ [v] = L^5, \tag{32}$$

$$\text{Proca}: \ [\mu] = \frac{1}{L}, \ [\lambda] = L^0, \ [v] = L^2.$$

Turning now to scaling symmetries, we notice the existence of *three* different transformations, which, in all cases, leave invariant the equations of motion (in the relations below, the functions or constants that are not mentioned explicitly remain invariant).

The first one (s0) is very simple

$$(s0): \quad \sigma \to c\sigma, \ w \to cw, \tag{33}$$

with c some arbitrary positive constant. However, this invariance is fixed when imposing the asymptotic flatness condition $\sigma \to 1$ as $r \to \infty$.

More importantly, the equations of motion are invariant under a scaling of the radial coordinate, together with other functions and parameters of the model. In the scalar and Proca cases, this transformation reads:

$$(s1): \quad \text{scalar and Proca}: \ r = c\bar{r}, \ w = \frac{1}{c}\bar{w}, \ \mu = \frac{1}{c}\bar{\mu}, \ \lambda = \frac{\bar{\lambda}}{c^2}, \ v = \frac{\bar{v}}{c^4}, \ m = c\bar{m}. \tag{34}$$

In the Dirac case, the corresponding symmetry is more complicated, with

$$(s1): \quad \text{Dirac}: \ r = c\bar{r}, \ w = \frac{1}{c}\bar{w}, \ \mu = \frac{1}{c}\bar{\mu}, \ v = c\bar{v}; \ m = c\bar{m}, \ f = \frac{\bar{f}}{\sqrt{c}}, \ g = \frac{\bar{g}}{\sqrt{c}}. \tag{35}$$

In all cases, the product $m\mu$ and the ratio w/μ are left invariant by the symmetry $(s1)$. Under this transformation, the global quantities behave as

$$M = c\bar{M}, \ Q = c^2\bar{Q}. \tag{36}$$

The symmetry $(s1)$ is usually employed in order to work in units of length set by the field mass,

$$\bar{\mu} = 1, \ i.e. \ c = \frac{1}{\mu}. \tag{37}$$

Finally, the equations are also invariant under a suitable scaling of the matter field(s) functions, together with some coupling constants (while the radial coordinate or the field frequency are not affected):

$$(s2): \ \lambda = \frac{\bar{\lambda}}{c^2}, \ v = \frac{\bar{v}}{c^2}, \ G = \frac{\bar{G}}{c^2}, \tag{38}$$

together with

$$(s2): \quad \text{scalar}: \ \phi = \bar{\phi}c, \quad \text{Dirac}: \ f = \bar{f}c, \ g = \bar{g}c, \quad \text{Proca}: \ F = \bar{F}c, \ H = \bar{H}c, \tag{39}$$

while the global quantities transform as

$$M = c^2\bar{M}, \ Q = c^2\bar{Q}. \tag{40}$$

The symmetries ($s1$) and ($s2$) are used, in practice, in order to simplify the numerical study of the solitons. First, the symmetry ($s2$) can be used to set $\bar{G} = 1$; i.e., essentially one absorbs Newton's constant in the expression of the matter field(s). This is the usual approach for the models without self-interaction, see, e.g., the discussion in [12].

However, the probe limit of the solutions becomes less transparent for this choice. An alternative route, employed in this work, is to use (38) and (39) to set the coefficient of the quartic term to unity,

$$\bar{\lambda} = 1, \quad i.e. \ c = \frac{1}{\sqrt{\lambda}}. \tag{41}$$

It follows that two mass scales naturally emerge, one set by gravity $M_{\mathrm{Pl}} = 1/\sqrt{G}$ and the other one, M_0, set by the field(s) coupling constants, with

$$\text{scalar and Proca}: \ M_0 = \frac{\mu}{\sqrt{\lambda}}; \quad \text{Dirac}: \ M_0 = \frac{1}{\sqrt{\lambda}}. \tag{42}$$

The ratio of these fundamental mass scales defines the dimensionless coupling constant

$$\alpha = \frac{M_0}{M_{\mathrm{Pl}}}, \tag{43}$$

which is relevant in the physics of the solutions.

Another dimensionless input parameter is the scaled constant for the sextic term in the potential U, with

$$\text{scalar and Proca}: \ \beta = \frac{\nu\mu^2}{\lambda^2}; \quad \text{Dirac}: \ \beta = \frac{\nu\mu}{\lambda^2}. \tag{44}$$

As such, the mass and charge of the non-gravitating solutions is given in units set by μ, λ, while, in the gravitating case, in order to make contact with the previous results without self-interaction, we use units that are set by G and μ.

A priori, the range of α is unbounded, $0 \leqslant \alpha < \infty$. The limit $\alpha \to 0$ corresponds to $G \to 0$, i.e., the probe limit—one solves the matter field(s) equations on a fixed geometry, which should be a solution of the vacuum Einstein equations. The limit $\alpha \to \infty$ corresponds to $\lambda \to 0$, i.e., no quartic, or sextic term (when using β, if ν is finite) in the action. Thus, solutions of the Einstein-matter field equations without self-interaction are approached in the second limit.

This choice of units with $\mu = \lambda = 1$ (after employing the above scaling symmetries) has the advantage of greatly simplifying the numerical study of the solutions. For example, the Einstein equations read

$$R_{\alpha\beta} - \frac{1}{2}g_{\alpha\beta} = 2\alpha^2 T_{\alpha\beta}, \tag{45}$$

the only input parameters being

$$\{\alpha, \ \beta \ \text{and} \ w\}, \tag{46}$$

with w the scaled frequency. Additionally, the scaled scalar potential reads

$$U = \psi^2 - \psi^4 + \beta\psi^6, \tag{47}$$

For $\psi^2 > 0$, U is strictly positive for $\beta > 1/4$. The case $\beta = 1/4$ is special, since the potential becomes $U = \psi^2(1 - \psi^2/2)^2$, and, thus, possesses three degenerate minima, at $\psi = \{0, \pm\sqrt{2}\}$. A discussion of these aspects (for a spin-zero field) can be found in Ref. [32].

3. The Probe Limit: Flat Spacetime Solutions

3.1. Deser's Argument and Virial-Type Identities

Before performing a numerical study of the solutions, it is useful to derive virial-type identities. For a flat spacetime background, we can adapt a simple argument given long ago by Deser [26] (used therein to rule out the existence of finite energy time-independent solutions in Yang–Mills theory) in order to obtain virial-type identities and a simple relation between the mass of solutions and the trace of the energy-momentum tensor (the same virial identities are found by adapting Derrick's scaling argument [33].)—see also [34].

Working in Cartesian coordinates x^a ($a = 1, 2, 3$), assume the existence of a stationary soliton in some field theory model. Following [26], consider the following (trivial) identity observe that Deser's argument cannot be extended to a curved spacetime background)

$$\frac{\partial}{\partial x^a}\left(x^b T_b^a\right) = T_a^a + x^b \frac{\partial T_b^a}{\partial x^a} \, , \tag{48}$$

together with its volume integral. The left hand side vanishes from regularity and finite energy requirements (note that, in the spin-1/2 case, one considers the total energy-momentum tensor). The second term in (48) vanishes from energy-momentum conservation (plus staticity) and, thus, we are left with the virial-type identity

$$\int d^3x \, T_a^a = 0. \tag{49}$$

It follows that the total mass-energy of a static soliton in $d = 3 + 1$ dimensions is determined by the trace of the energy-momentum tensor

$$M = -\int d^3x \, T_t^t = -\int d^3x \, T_\mu^\mu \, . \tag{50}$$

When applied to the specific Ansätze in this work, (49) results in the following expressions:

- scalar field:

$$\int_0^\infty dr \, r^2 \left[\frac{1}{3}\phi'^2 + (\mu^2 - w^2)\phi^2 + v\phi^6\right] = \lambda \int_0^\infty dr \, r^2\phi^4. \tag{51}$$

This relation can be simplified by using the equation for ϕ in a simpler form

$$\int_0^\infty dr \, r^2(\mu^2 - w^2)\phi^2 = \frac{\lambda}{2} \int_0^\infty dr \, r^2\phi^4. \tag{52}$$

This clearly shows that the (flat space) Q-ball solutions are supported by the quartic self-interacting term, with $\lambda > 0$ (i.e., the sextic term is not relevant at this level).

- Dirac field:

$$\int_0^\infty dr \, r^2 \left(gf' - fg' + \frac{2fg}{r} + \frac{3}{8}U - \frac{3}{2}w(f^2 + g^2)\right) = 0 \tag{53}$$

which can also be written, via field equations, in the alternative form

$$\int_0^\infty dr \, r^2 \left(\mu\psi^2 + \lambda\psi^4\right) = \int_0^\infty dr \, r^2 \left(4w(f^2 + g^2) + 3v\psi^6\right). \tag{54}$$

One concludes that, for $\nu = 0$ and $\lambda > 0$, the solutions are supported by the harmonic time dependence. However, the above relation does not clarify the role that is played by the nonlinear quartic term for the existence of solitons.

- Proca field:

$$\int_0^\infty dr\, r^2 \left(\frac{1}{2}(F' - wG)^2 - 3U + 2G^2 \frac{dU}{d\mathcal{A}^2} \right) = 0 \,. \tag{55}$$

After eliminating the kinetic term $(F' - wG)^2$, via field equations, the above relation takes the suggestive form:

$$\int_0^\infty dr\, r^2 \left[\mu^2(3F^2 + (\frac{\mu^2}{w^2} - 1)G^2 + \frac{16\lambda^2 G^2}{w^2}\mathcal{A}^4 + \frac{6\nu}{w^2}\mathcal{A}^4 \left(w^2(F^2 + G^2) + 2G^2(\mu^2 + 3\nu\mathcal{A}^2) \right) \right] = $$
$$2\lambda \int_0^\infty dr\, r^2 \mathcal{A}^2 \left(3F^2 + G^2 + \frac{4G^2}{w^2}(\mu^2 + 6\nu\mathcal{A}^2) \right) . \tag{56}$$

From the bound state condition $\mu^2 \geqslant w^2$, it is clear that the existence of finite mass solutions requires a quartic term, $\lambda \neq 0$ (ν being irelevant). However, since $\mathcal{A}^2$ may take both positive or negative values, one cannot use the above relation in order to predict the sign of λ.

It is also interesting to note that the mass of the flat space Proca solitons takes the simple form

$$M = \int_0^\infty dr\, r^2 \left(\mu^2 \mathcal{A}^2 - 2\nu \mathcal{A}^6 \right) . \tag{57}$$

3.2. Numerical Results

3.2.1. General Remarks

The corresponding equations are found by taking

$$N = \sigma = 1 \,, \tag{58}$$

in the corresponding general equations presented in Section 2, and we shall not display them here. Moreover, the boundary conditions that are satisfied by the functions ψ at $r = 0, \infty$ are similar to those in the gravitating case, as given in the next Section.

The case of a scalar field is special for a flat spacetime metric. The frequency parameter w is not relevant, since w^2 acts as an effective *tachyonic* contribution to the mass term and, thus, it can be absorbed into μ^2, by defining $\mu^2 - w^2 \to \mu^2$. After this redefinition, the scalar field is static, $\Phi = \phi(r)$ and, thus, the Noether charge vanishes. Therefore, all Q-ball solutions in a flat spacetime background can be interpreted as *static* scalar solitons, in a model with a shifted scalar field mass term [35], for a new potential

$$U = U_{(w)} - w^2\phi^2. \tag{59}$$

Note that, although ϕ formally satisfies the same equation as before, the energy-momentum tensor and the total mass of these solutions are different. Additionally, the virial identities imply that the redefined potential U is necessarily negative for some range of ϕ, which is realised by the solutions, $U < 0$. Moreover, one can prove the following relation [35]

$$M(w = 0) = M(w) - wQ, \tag{60}$$

which relates the mass of a static solution ($w = 0$ and the redefined potential (59)) to the mass and Noether charge of solutions with a given w (the other parameters in the potential are kept fixed).

No similar relation seems to exist for higher spin fields. However, it is interesting to notice the existence in this case of a curious static, purely electric solution, i.e., with $H = 0, w = 0$. The electric potential $F(r)$ essentially satisfies the same equation as a scalar Q-ball,

$$F'' + \frac{2F'}{r} - (\mu^2 + 4\lambda F^2 + 6\nu F^4)F = 0 . \tag{61}$$

The existence of solutions with proper asymptotics requires $\lambda < 0$, being constructed within the same scheme as in the generic case.

To the best of our knowledge, this case has not been discussed in the literature. However, they possess some unphysical features. For example, by using the virial identity (57) together with the above equations, one can prove that the total mass of this configuration is negative (moreover, we have verified that M remains negative when including gravity effects)

$$M = -\frac{1}{3} \int_0^\infty dr r^2 F'^2 < 0. \tag{62}$$

In all cases, the sextic self-interaction term is not necessary for the existence of solutions. Because the $\beta = 0$ case has some special properties, we shall discuss it separately at the end of this section.

3.2.2. Solutions with a Sextic Self-Interaction Term, $\beta > 0$

In what follows, in order to simplify the picture, we shall assume $\beta > 1/4$, such that the potential $U(\psi^2)$ (as given by (6)) is strictly positive (assuming $\psi^2 > 0$, which, as we shall see, is not necessarily the case).

Because no exact solutions are known, the solitons are constructed numerically. Figures 1–3 (left panels) show the profile of typical solutions. One notices that, in the Proca case, $\psi^2 = \mathcal{A}^2$ is negative for small enough r, while $\psi^2 = i\overline{\Psi}^{(A)}\Psi^{(A)}$ is positive in the Dirac case.

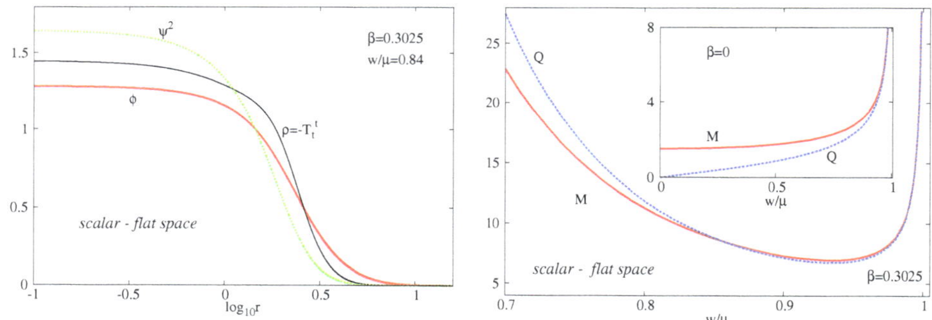

Figure 1. (**Left**): the radial profile of a typical non-gravitating scalar soliton. (**Right**): the mass and Noether charge are shown *vs.* the scalar field frequency for the fundamental family of non-gravitating scalar solitons.

The numerical results indicate that the $s = 0, 1/2, 1$ flat spacetime solitons follow the same pattern, which can be summarized, as follows. First, in all cases, the solutions only exist in a certain frequency range, $w_{\min} < w < w_{\max} = \mu$, with $w_{\min}$ determined by β. Second, the solutions with $w < \mu$ decay exponentially in the far field, with no radiation, as $\psi \sim e^{-\sqrt{\mu^2 - w^2}r}$. Finally, the most interesting qualitative feature is perhaps the existence in all cases of a mass gap. That is, at a critical value of the frequency, both the mass and charge of the solutions assume their minimal (nonzero) value, from where they monotonically increase towards both limiting values of the frequency (see Figures 1–3 (right panels)). When considering the mass of the solutions as a function of the charge, there are thus two branches, merging at the minimal charge/mass. One expects that a subset of solutions with M smaller than the mass of Q free particles (bosons or fermions) may be stable, which is possible along the lower frequency branch.

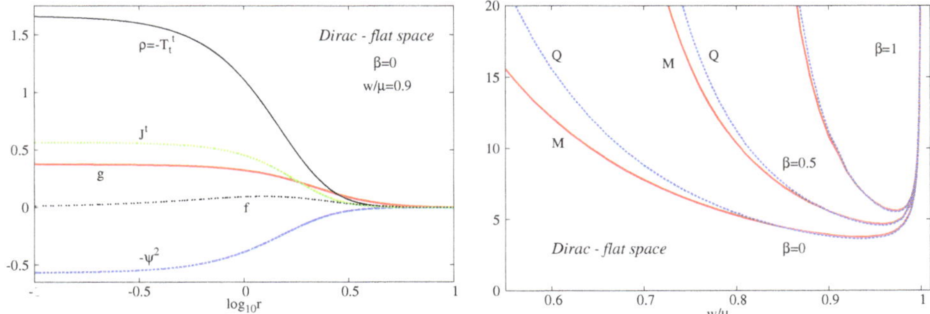

Figure 2. Same as Figure 1 for Dirac stars. Note that the single particle condition, $Q = 1$, is not imposed here.

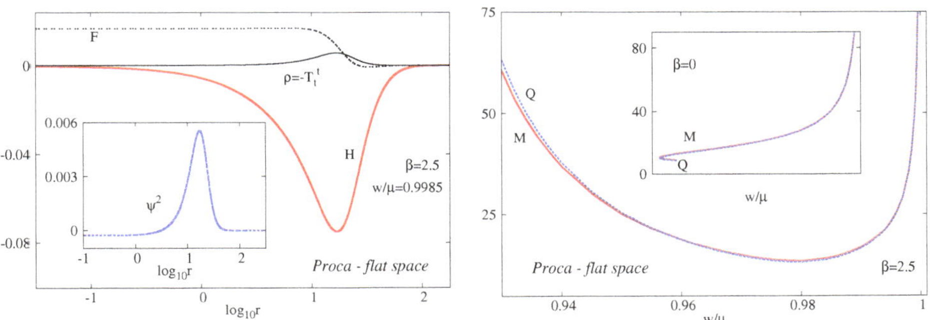

Figure 3. Same as Figure 1 for Proca stars.

3.2.3. Solutions without a Sextic Self-Interaction Term, $\beta = 0$

This case is likely physically less relevant (since the potential U is not positive definite); however, it possesses some interesting properties, which depend, to some extent, on the spin of the field.

Starting with the scalar case, some numerical results are displayed in Figure 1 (right panel, inset). Note that all of these solutions are unstable, since $M > Q$, and, thus, they have excess energy. We remark that solutions with a negative energy density region, $\rho < 0$, found on a small $r-$interval, are found for small enough w (although $M > 0$ always).

In fact, the equation for the scalar field has an interesting form, with the frequency parameter being irrelevant. After using the alternative rescaling

$$r \rightarrow r/\sqrt{\mu^2 - w^2}, \quad \phi \rightarrow \phi\sqrt{\frac{\mu^2 - w^2}{\lambda}}, \tag{63}$$

one can see that no free parameter exists in this case. The scalar field satisfies the kink-like equation

$$\frac{1}{r^2}\frac{d}{dr}\left(r^2\frac{d\phi}{dr}\right) = \phi - 2\phi^3, \tag{64}$$

which does not seem to possess an exact solution (Equation (64) is discussed by many authors, being relevant for the issue of false vacuum decay [36]—e.g., an existence proof can be found in Ref. [37]). Additionally, one can show that the following relation holds

$$M(w; \mu, \lambda) = \frac{M(w = 0; \mu, \lambda)}{\sqrt{\mu^2 - w^2}}, \tag{65}$$

with $M(w = 0; 1, 1) \simeq 1.503$.

The pattern that is exhibited by the Proca solutions without a sextic term is different, as displayed in Figure 3. As found in Ref. [24], the limit $w = 0$ is not approached in this case. Instead, a minimal value of w is reached, with a backbending towards a critical solution possessing finite charges. Because $M > Q$, one expects all of these solutions to be unstable. Additionally, the mass is still positive, $M > 0$, and we did not notice the existence of negative energy densities (although we could not find an analytical argument to show that $\rho > 0$).

Finally, the pattern of Dirac solitons without a sextic self-interaction (which was the original case in the pioneering work by Soler [20]) seems to be similar to the one that is found in the $\beta > 0$ case, see Figure 2 (right panel). In particular, both M and Q still diverge at the limits of the w-interval.

4. Including the Gravity Effects

4.1. The Boundary Conditions

The boundary conditions satisfied by the metric functions at the origin are

$$m(0) = 0, \quad \sigma(0) = \sigma_0, \tag{66}$$

while at infinity, one imposes

$$m(\infty) = M, \quad \sigma(\infty) = 1, \tag{67}$$

with $\sigma(0)$, M numbers fixed by numerics. The matter functions should vanish as $r \to \infty$

$$\phi(\infty) = f(\infty) = g(\infty) = F(\infty) = H(\infty) = 0, \tag{68}$$

while the boundary conditions at the origin are

$$\begin{aligned} \text{scalar}: \left.\frac{d\phi(r)}{dr}\right|_{r=0} &= 0, \\ \text{Dirac}: f(0) = 0, \quad \left.\frac{dg(r)}{dr}\right|_{r=0} &= 0, \\ \text{Proca}: \left.\frac{dF(r)}{dr}\right|_{r=0} &= 0, \quad H(0) = 0. \end{aligned} \tag{69}$$

Let us mention that, in each case, one can construct an approximate form of the solutions both at $r = 0$ and at infinity, which is compatible with the boundary conditions above.

4.2. Virial Identities

The reduced action (25) allows for us to prove that the solutions satisfy the (simple enough) interesting virial identities, which generalize for a curved geometry Deser's relations (51), (53) and (55).

Following [38,39] (wherein generalizations for a curved geometry of Derrick's argument for flat spacetime [33] were established), assume the existence of a solution that is described by $m(r)$, $\sigma(r)$ and $\psi(r)$ (with $\psi = \{\phi; f, g; F, H\}$ the matter fields) with suitable boundary conditions at $r = 0$ and at infinity. Subsequently, each member of the one-parameter family $m_\Lambda(r) \equiv m(\Lambda r)$, $\sigma_\Lambda(r) \equiv \sigma(\Lambda r)$, and $\psi_\Lambda(r) \equiv \psi(\Lambda r)$, assumes the same boundary values at $r = 0$ and at $r = \infty$, and the action $S_\Lambda \equiv S_{eff}[m_\Lambda, \sigma_\Lambda, \psi_\Lambda]$ must have a critical point at $\Lambda = 1$, i.e., $[dS/d\Lambda]_{\Lambda=1} = 0$. Thus, the putative solution must satisfy the following virial relations (here, it is useful to work with unscaled variables, i.e., before considering the transformations (s2), (s3)):

- scalar field:

$$\int_0^\infty dr\, r^2 \sigma \left[\phi'^2 + \frac{w^2(1-4N)\phi^2}{N^2\sigma^2} + 3U \right] = 0 ; \tag{70}$$

- Dirac field:

$$\int_0^\infty dr\, r^2 \sigma \left[\sqrt{H}(gf' - fg')\left(2 + \frac{m}{rN}\right) + \frac{4fg}{r} - \frac{w(f^2+g^2)}{\sigma\sqrt{N}}\left(3 - \frac{m}{rN}\right) - 3U \right] = 0 ; \tag{71}$$

- Proca field:

$$\int_0^\infty dr\, r^2 \sigma \left[\mu^2 \left(\mathcal{A}^2 \left(4 - \frac{1}{N}\right) + 2H^2(1-N) \right) + 2\lambda \mathcal{A}^2 \left(\mathcal{A}^2 \left(5 - \frac{2}{N}\right) + 4H^2(1-N) \right) \right.$$
$$\left. + 6\nu \mathcal{A}^4 \left(\mathcal{A}^2 \left(2 - \frac{1}{N}\right) + 2H^2(1-N) + 4G^2(1-N) \right) - \frac{(wH - F')(3wH - F')}{\sigma^2} \right] = 0. \tag{72}$$

This relation can be further simplified after using the Proca field equations to yield

$$\int_0^\infty dr\, r^2 \sigma \left[\mu^2 \left(4 - \frac{1}{N}\right) \mathcal{A}^2 + 2\lambda \left(5 - \frac{2}{N}\right) \mathcal{A}^4 + 6\nu \left(2 - \frac{1}{N}\right) \mathcal{A}^6 \right] = \tag{73}$$
$$2 \int_0^\infty dr\, r^2 \sigma H^2 (\mu^2 + 4\lambda \mathcal{A}^2 + 6\nu \mathcal{A}^4) \left[2N - 1 + \frac{N^2\sigma^2}{2w^2}(\mu^2 + 4\lambda \mathcal{A}^2 + 6\nu \mathcal{A}^4) \right]$$

These expressions are not transparent, with the effects of gravity and self-interaction being mixed. As such, their main use is to test the accuracy of the numerical results. However, the situation changes once we set $\lambda = \nu = 0$ (i.e., no self-interaction). Subsequently, one can see that the solutions are supported by an harmonic time dependence of the matter fields, $w \neq 0$, except for the Dirac case, where we could not prove a similar result.

4.3. General Features

The flat space solitons can be generalized to curved spacetime. The presence of higher order self-interacting terms in the potential is not crucial for the existence of gravitating solutions. However, they affect some of their quantitative features.

A common pattern emerges again, with the basic generic properties of the gravitating solutions being summarized, as follows. First, the solutions are topologically trivial, with $0 \leqslant r < \infty$. They possess no horizon; the size of the S^2-sector of the metric shrinks to zero as $r \to 0$. At infinity, a Minkowski background is approached. Second, perhaps the most interesting feature is that gravity regularizes the divergencies of the mass and charge that are found in the probe case at the limit(s) of the w-interval. In particular, this regularization implies that no mass gap exists for gravitating configurations: $M, Q \to 0$ as $w \to \mu$. Additionally, assuming the existence of a sextic self-intertion term, with $\beta > 1/4$, in all cases, the family of solutions describes a continuous curve in a mass M (or charge Q), vs. frequency, w, diagram. This curve starts from $M = Q = 0$ for $w = \mu$, in which limit the fields become very diluted and the solution trivializes. At some intermediate frequency, a maximal mass is attained (that may be a global, or only local, maximum, depending on α, β).

The effects of the quartic and sextic self-interacting terms in the potential (6) become irrelevant for large enough α. For example, the $\alpha = 10$ curves that are displayed in Figures 4–6 are well approximated by the corresponding Einstein-scalar/Dirac/Proca results, with $\beta = \nu = 0$, the maximal relative difference being of only a few percent (towards the critical value of frequency).

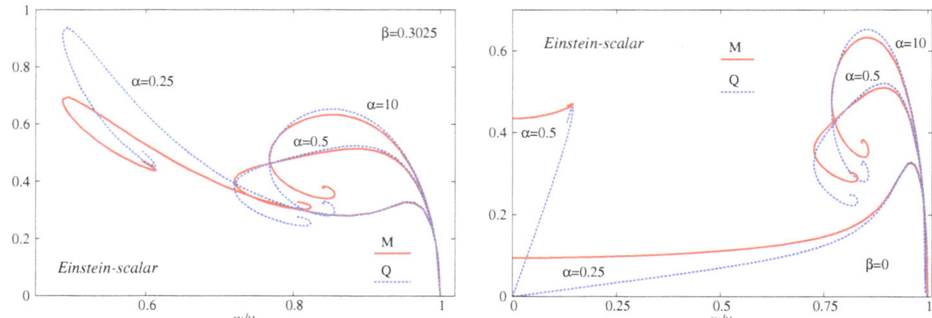

Figure 4. ADM mass and Noether charge of the gravitating scalar boson stars *vs.* the scalar field frequency for families of solutions with three different values of the coupling constant α. The solutions in the right panel do not possess a sextic self-interacting term.

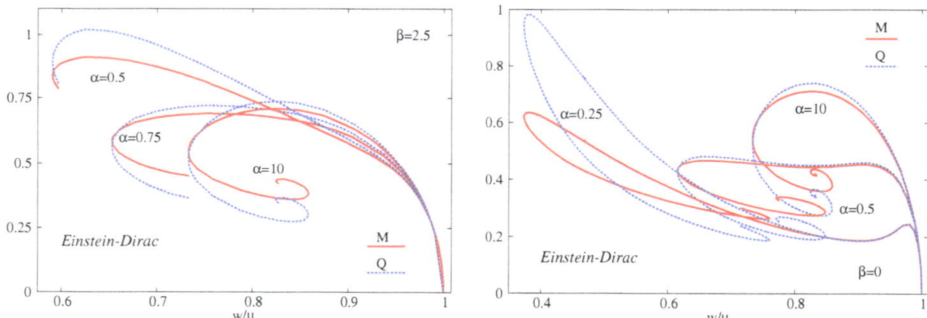

Figure 5. The same as Figure 4 for Dirac stars. The single particle condition, $Q = 1$, is not imposed here.

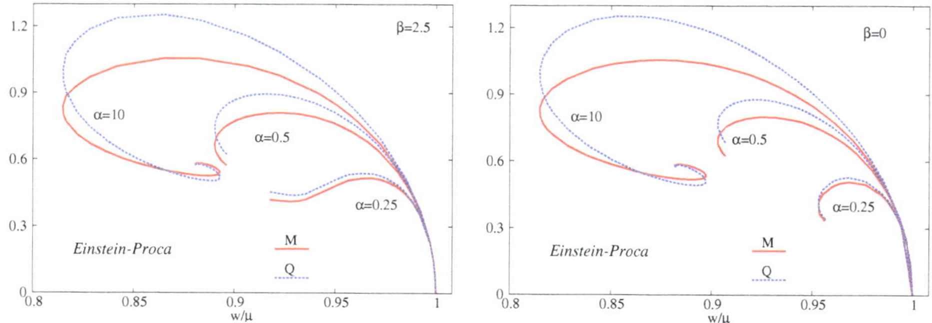

Figure 6. Same as Figure 4 for Proca stars.

One can also identify some specific features, as follows. The scalar solutions with $\beta > 1/4$ (i.e., a positive potential, $U > 0$) and the Proca starts with large α describe a spiral in a (M, w)-diagram.

As in similar cases, likely, these spirals approach, at their centre, a critical singular solution. The Dirac solutions with $\beta = 0$ also describe a spiral. For general Proca solutions with $\beta = 0$ and Dirac solutions with $\beta \neq 0$, the (M, w)-curves appear to end in a critical configuration before describing a spiral. In the Proca case, this feature is discussed in [24].

The scalar field solutions with $\beta = 0$ possess a more complicated pattern. For small α, they possess a static limit. Moreover, two disconnected branches of solutions exist for some intermediate range of α e.g., $\alpha = 0.5$ in Figure 4 (right panel). In addition to the familiar spiral starting at $w/\mu = 1$ and ending for some critical nonzero $w = w_c$, one finds a secondary set, which extends from $w = 0$ to some maximal value of $w < w_c$.

Finally, let us mention the existence of another interesting possibility, with quartic interaction only and $\lambda < 0$, in which case no flat spacetime solitons are found. The corresponding solutions were discussed in [40] for a scalar field, and in [25] for a Proca field. No similar study exists so far for a fermionic field. In the bosonic case, perhaps the most interesting feature is that their maximal mass is proportional with $\sqrt{|\lambda|}$, and, thus, can increase dramatically.

5. Other Aspects

5.1. No Hair Results

It is interesting to inquire whether the solutions above may allow the existence of a black hole horizon, inside. Indeed, this is a well known feature that is found for a variety of other solitons, see e.g., the review work [27]. However, this is not the case for the (relatively) simpler solutions in this work.

The situation can be summarized, as follows:

- **Scalar case**
  Peña and Sudarsky established a no-scalar-hair theorem, ruling out a class of spherically symmetric BHs with scalar hair [41]. Their proof also covers the case of the potential (6) considered in this work, and it still holds if the hair has the harmonic time-dependence that we consider.
- **Dirac case**
  No hair results for the Einstein–Dirac case with a massive spinor (no self-interaction) were proposed in [42,43]. The nonexistence of stationary states for the nonlinear Dirac equation with a quartic self-interaction on the Schwarzschild metric has been proven in [44].
- **Proca case**
  A no hair theorem has been proven in [30] for a massive, non-selfinterating Proca field. In Appendix B, we generalize it for an arbitrary Proca potential $U(\mathcal{A}^2)$.

5.2. The Issue of Particle Numbers: Bosons vs. Fermions

In all of the results displayed above, we have equally treated the bosonic and fermionic fields. However, while the classical treatment of the Dirac equation is mathematically justified (a discussion on classical spinors and their possible physical justification can be found, e.g., in Ref. [45]), physically, its fermionic nature should be imposed at the level of the occupation number: at most, a single, particle, in accordance to Pauli's exclusion principle.

Similarly to the non-self-interacting case that is discussed in [12], the one particle condition is imposed, as follows. Let us suppose that we have a numerical solution with some values for the mass and charge $(M_{(\mathrm{num})}, Q_{(\mathrm{num})})$. Subsequently, we can use the symmetry (34)–(36) in order to map it to a 'new' solution with

$$\bar{Q}_{(\mathrm{num})} = Q_0, \tag{74}$$

where $Q_0 > 0$ is arbitrary. This assumption fixes the value of the scaling parameter c,

$$c = \sqrt{\frac{Q_{(\mathrm{num})}}{Q_0}}, \tag{75}$$

such that the *numerical* mass of the 'new' solution will be [12]

$$\bar{M}_{(\text{num})} = M_{(\text{num})} \sqrt{\frac{Q_0}{Q_{(\text{num})}}} \, . \tag{76}$$

Note that the values of w, μ, λ, and ν should also be scaled according to (34) and (35).

This transformation is used to normalize the total charge of a fermion to one, $Q_0 = 1$. With this condition $Q = 1$, the (M, w)-curves in Figure 2 (right panels) and Figure 5 are not sequences of solutions with fixed parameters in the potential and varying Q; instead, they get mapped into sequences with fixed Q and varying μ, ν. Consequently, one is discussing a sequence of solution of different models (since μ, ν are input parameters in the action).

Figure 7 shows the corresponding results, for both the probe limit case and including gravity effects. An interesting feature here is that the mass of $Q = 1$ non-gravitating configuration can still take arbitrary large values. However, as expected, gravity effects lead to a picture that is qualitatively similar to that found in the $\lambda = \nu = 0$ case [12]. Again, both the total mass, M and the mass of the field μ are bounded and never exceed, roughly, M_{Pl}.

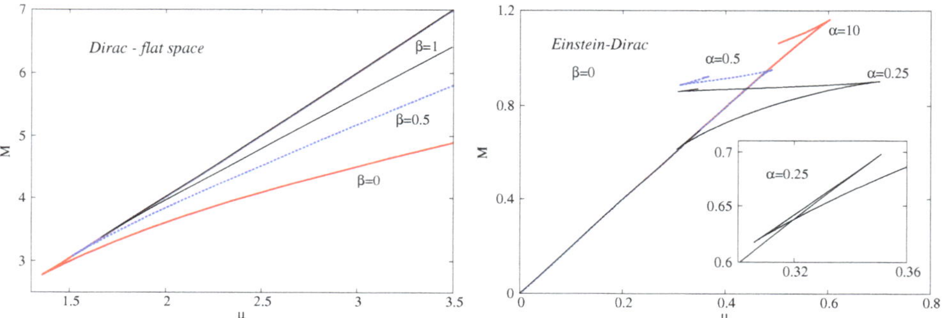

Figure 7. (**Left panel**) Soliton mass *vs.* the mass of the elementary quanta of the field, for non-gravitating solutions of the Dirac equations with three different values of β. (**Right panel**) The same for the gravitating solutions with several values of the coupling constant α and $\beta = 0$. The single particle condition, $Q = 1$, is imposed here.

6. Further Remarks. Conclusions

The main purpose of this work was to provide a comparative analysis of three different types of solitonic solutions of GR coupled with matter fields of spin $s = 0, 1/2, 1$. A unified framework has been proposed, analysing these cases side-by-side under a consistent set of notations and conventions. Differently from the previous work [12], the matter fields herein are self-interacting, such that all three models possess (Q-ball-like) solutions on the flat spacetime limit. As such, a more complicated landscape of gravitating solutions is found, with some new qualitative features when compared to the picture that is revealed in [12].

However, despite this fact, our study shows that there is, again, a certain universality in the properties of the solitons, being to some extent independent of their (fundamental) spin. As with linear matter fields, the basic ingredients are again: (**i**) a harmonic time dependence of the matter fields; (**ii**) complex field(s)/multiplets such that the energy–momentum is still real; and, (**iii**) the existence of of mass term as a trapping mechanism, creating bound states with $w < \mu$. Additionally, if one requires the presence of a well defined flat space limit, then (**iv**) the fields should possess at least a quartic self-interaction term.

As an avenue for future work, it would be interesting to go beyond the case of spherical symmetry and consider a comparative study of axially symmetric, spinning solutions. In the non-self-interacting case, this was the subject of the recent work [46], where a common pattern was again revealed to exist. The situation in the presence of self-interactions is less studied; only the scalar field case has, so far, been discussed in the literature [47,48]. Indeed, even flat space spinning solitons with spin $s = 1/2, 1$ are yet unreported in the literature. Moreover, even in the static case, new families of non-spherically symmetric solitons should exist, generalizing, for a self-interacting potential (and possibly for a higher spin), the $s = 0$ multipolar boson stars recently reported in [49].

Finally, let us remark that, for a bosonic field, $s = 0, 1$, the no-hair theorems (as reviewed in the previous Section) can be circumvented for an horizon that rotates synchronously with the field, leading to BHs with scalar or Proca hair [30,50]. However, this does not seem to be the case for fermions, regardless of the presence of a self-interacting potential.

Author Contributions: The two authors contributed equally to the conceptualization, methodology, software, validation, formal analysis, investigation, resources, data curation, writing–original draft preparation, writing–review and editing, visualization, supervision, project administration, funding acquisition. All authors have read and agreed to the published version of the manuscript.

Funding: This work is supported by the Center for Research and Development in Mathematics and Applications (CIDMA) through the Portuguese Foundation for Science and Technology (FCT - Fundacao para a Ciência e a Tecnologia), references UIDB/04106/2020 and UIDP/04106/2020 and by national funds (OE), through FCT, I.P., in the scope of the framework contract foreseen in the numbers 4, 5 and 6 of the article 23, of the Decree-Law 57/2016, of August 29, changed by Law 57/2017, of July 19. We acknowledge support from the projects PTDC/FIS-OUT/28407/2017, CERN/FIS-PAR/0027/2019 and PTDC/FIS-AST/3041/2020. This work has further been supported by the European Union's Horizon 2020 research and innovation (RISE) programme H2020-MSCA-RISE-2017 Grant No. FunFiCO-777740. The authors would like to acknowledge networking support by the COST Action CA16104. Computations were performed at the Blafis cluster, in Aveiro University.

Conflicts of Interest: The authors declare no conflict of interest.

Appendix A. The Dirac Field: Conventions

Since this case is more complicated, we shall include here the basic relations.

Following the framework in [31], we consider a general four dimensional metric $g_{\alpha\beta}$, and introduce a tetrad of vectors

$$e_a^\alpha = \{e_0^\alpha, e_1^\alpha, e_2^\alpha, e_3^\alpha\}, \tag{A1}$$

which we take to be an orthonormal basis, i.e.,

$$g_{\alpha\beta} e_a^\alpha e_b^\beta = \eta_{\alpha\beta}, \tag{A2}$$

where

$$\eta_{ab} = \mathrm{diag}(-1, 1, 1, 1). \tag{A3}$$

Here, Roman and Greek letters are used for tetrad and coordinate indices, respectively. Roman indices are raised and lowered with η_{ab}. It follows that

$$e_\alpha^a = \eta^{ab} g_{\alpha\beta} e_b^\beta \quad \text{and} \quad g_{\alpha\beta} = \eta_{ab} e_\alpha^a e_\beta^b. \tag{A4}$$

The next step is to define two sets of 4×4 matrices γ^α and $\hat\gamma^a$ satisfying the anticommutation relations

$$\{\gamma^\alpha, \gamma^\beta\} = 2g^{\alpha\beta} I_4, \quad \{\hat\gamma^a, \hat\gamma^b\} = 2\eta^{ab} I_4, \tag{A5}$$

(where, as usual, $\{A, B\} = AB + BA$). Note that the former set γ^α are functions of spacetime position, whereas the latter $\hat{\gamma}^a$ have constant components. The two sets may be related with any orthonormal tetrad,

$$\gamma^\alpha = e^\alpha_a \hat{\gamma}^a. \tag{A6}$$

Matrix indices are raised/lowered in the standard way: $\hat{\gamma}_a = \eta_{ab}\hat{\gamma}^b$ and $\gamma_\alpha = g_{\alpha\beta}\gamma^\beta$.

One uses the Weyl/chiral representation, in which

$$\tilde{\gamma}^0 = \begin{pmatrix} O & I \\ I & O \end{pmatrix}, \quad \tilde{\gamma}^i = \begin{pmatrix} O & \sigma_i \\ -\sigma_i & O \end{pmatrix}, \quad i = 1, 2, 3, \tag{A7}$$

where I is the 2×2 identity, O is the 2×2 zero matrix, and σ_i are the Pauli matrices

$$\sigma_1 = \begin{pmatrix} 0 & 1 \\ 1 & 0 \end{pmatrix}, \quad \sigma_2 = \begin{pmatrix} 0 & -i \\ i & 0 \end{pmatrix}, \quad \sigma_3 = \begin{pmatrix} 1 & 0 \\ 0 & -1 \end{pmatrix}. \tag{A8}$$

Then the matrices $\hat{\gamma}^a$ are defined as

$$\hat{\gamma}^1 = i\tilde{\gamma}^3, \ \hat{\gamma}^2 = i\tilde{\gamma}^1, \ \hat{\gamma}^3 = i\tilde{\gamma}^2, \ \hat{\gamma}^0 = i\tilde{\gamma}^0. \tag{A9}$$

The Dirac four-spinor is written as

$$\Psi = \begin{pmatrix} \psi_- \\ \psi_+ \end{pmatrix}, \tag{A10}$$

where ψ_+ and ψ_- are (left- and right-handed) two-spinors, which may be projected out from Ψ with the operators $P_\pm = \frac{1}{2}\left(I \pm \hat{\gamma}_5\right)$ where

$$\hat{\gamma}^5 = i\hat{\gamma}^0\hat{\gamma}^1\hat{\gamma}^2\hat{\gamma}^3 = \begin{pmatrix} -I & O \\ O & I \end{pmatrix}. \tag{A11}$$

We define also the Dirac conjugate

$$\overline{\Psi} \equiv \Psi^\dagger \alpha, \tag{A12}$$

with $\Psi^\dagger$ denoting the usual Hermitian conjugate and $\alpha = -\hat{\gamma}^0$.

The spinor covariant derivative $\hat{D}_\nu$ which enters the Dirac Equation (11) is

$$\hat{D}_\nu = \partial_\nu - \Gamma_\nu. \tag{A13}$$

The spinor connection matrices Γ_ν are defined, up to an additive multiple of the unit matrix, by the relation

$$\partial_\nu \gamma^\mu + \Gamma^\mu_{\ \nu\lambda}\gamma^\lambda - \Gamma_\nu \gamma^\mu + \gamma^\mu \Gamma_\nu = 0, \tag{A14}$$

where $\Gamma^\mu_{\ \nu\lambda}$ is the affine connection. A suitable choice satisfying (A14) makes use of the spin connection $\omega_{\alpha\,bc}$,

$$\Gamma_\alpha = -\frac{1}{4}\omega_{\alpha\,bc}\hat{\gamma}^b\hat{\gamma}^c, \tag{A15}$$

the spin-connection $\omega_{\mu ab}$ being defined as

$$\omega_\mu{}^a{}_b = e^a_\nu e^\lambda_b \Gamma^\nu_{\ \mu\lambda} - e^\lambda_b \partial_\mu e^a_\lambda. \tag{A16}$$

Also, the covariant derivative of the conjugate spinor is

$$\hat{D}_\mu \overline{\Psi} = \partial_\mu \overline{\Psi} + \overline{\Psi}\Gamma_\mu. \tag{A17}$$

Appendix B. Self-Interacting Proca Field: A No Hair Result

As for the previous study [30] for a non-self-interacting field, the theorem is established by contradiction. Let us assume the existence of a regular BH solution of the Einstein-Proca equations. The general Ansatz and the field equations derived in Section 2 apply also to this case. Differently from the globally regular case, the geometry would possess a non-extremal horizon at, say, $r = r_H > 0$, which requires that

$$N(r_H) = 0\,, \tag{A18}$$

while $\sigma(r_H) > 0$. Since we are assuming that there are no more exterior horizons, then $r > r_H =$ constant are timelike surfaces and $N'(r_H) > 0$. Also, we can choose without loss of generality that $\sigma(r_H) > 0$, since the equations of motion are invariant under $\sigma \to -\sigma$. It follows that $N(r)$ and $\sigma(r)$ are strictly positive functions for any $r > r_H$.

In establishing the theorem, we shall use the relation (12) written in the generic form

$$\frac{d}{dr}\left(r^2 N\sigma H\dot{U}\right) = -\frac{wr^2 F\dot{U}}{N\sigma}, \tag{A19}$$

together with one of the Proca equations

$$F' = wH\left(1 - \frac{2N\sigma^2\dot{U}}{w^2}\right). \tag{A20}$$

The regularity of the horizon implies that the energy density of the Proca field is finite there and also the norm of the Proca-potential $\mathcal{A}$. One can easily see that this condition implies

$$F(r_H) = 0\,. \tag{A21}$$

Then the function $F(r)$ starts from zero at the horizon and remains strictly positive (or negative) for some r-interval (the case of a negative $F'(r)$ can be discussed in a similar way). Now, let us assume $F'(r) > 0$ for $r_H < r < r_1$. It follows that, in this interval, $F(r)$ is a strictly increasing (and positive) function.

Next, we consider the expression (which appears in Equations (A19) and (A20)),

$$P(r) \equiv 1 - \frac{2\sigma^2(r)N(r)\dot{U}}{w^2}\,. \tag{A22}$$

One can see that $P(r_H) = 1$; actually P becomes negative for large r, since $N \to 1$, $\sigma \to 1$ as $r \to \infty$, while $\mathcal{A}^2 \to 0$ and $\mu > w$ (which is a bound state condition necessary for an exponential decay of the Proca field at infinity). But the important point is the existence of an r-interval $r_H < r < r_2$ where $P(r)$ is a strictly positive function. Now, the same reasoning applies also for $\dot{U}$ (since $\dot{U}(r_H) = \mu^2/2 > 0$), while $\dot{U} \to 0$ asymptotically. This implies the existence of some interval in the vicinity of the horizon where $\dot{U}$, F and P are all positive.

At this point, let us consider an arbitrary value of r in this interval. Then we observe that (A19) implies

$$r^2\sigma(r)N(r)H(r)\dot{U} = -w\int_{r_H}^{r} dx\,\frac{x^2}{\sigma(x)N(x)}F(x)\dot{U}(x) < 0 \tag{A23}$$

in that interval. Consequently, $H(r) < 0$, since σ, N are positive everywhere outside the horizon.

The last conclusion implies a contradiction: $H(r) < 0$ is not compatible with $F'(r) > 0$, in that interval. In fact, $F'(r) > 0$ together with $P > 0$ and $w > 0$, from (A20), that $H(r) > 0$. Thus we conclude that $F(r) = H(r) = 0$ is the only solution compatible with a BH geometry (*q.e.d.*).

References

1. Thomson, W. On Vortex Atoms. *Proc. R. Soc. Edinb.* **1867**, *6*, 94–105. [CrossRef]
2. Skyrme, T.H.R. A Nonlinear field theory. *Proc. R. Soc. Lond. A* **1961**, *260*, 127–138.
3. Skyrme, T.H.R. A Unified Field Theory Of Mesons And Baryons. *Nucl. Phys.* **1962**, *31*, 556–569. [CrossRef]
4. Rosen, N. A Field Theory of Elementary Particles. *Phys. Rev.* **1939**, *55*, 94. [CrossRef]
5. Klinkhamer, F.R.; Manton, N.S. A Saddle Point Solution in the Weinberg-Salam Theory. *Phys. Rev. D* **1984**, *30*, 2212. [CrossRef]
6. Hooft, G. Magnetic Monopoles in Unified Gauge Theories. *Nucl. Phys. B* **1974**, *79*, 276–284. [CrossRef]
7. Polyakov, A.M. Particle Spectrum in the Quantum Field Theory. *JETP Lett.* **1974**, *20*, 194–195.
8. Manton, N.S.; Sutcliffe, P. *Topological Solitons*; Cambridge University Press: Cambridge, UK, 2004.
9. Shnir, Y.M. *Topological and Non-Topological Solitons in Scalar Field Theories*; Cambridge University Press: Cambridge, UK, 2018.
10. Lee, T.D.; Pang, Y. Nontopological solitons. *Phys. Rept.* **1992**, *221*, 251–350. [CrossRef]
11. Radu, E.; Volkov, M.S. Existence of stationary, non-radiating ring solitons in field theory: knots and vortons. *Phys. Rept.* **2008**, *468*, 101. [CrossRef]
12. Herdeiro, C.A.R.; Pombo, A.M.; Radu, E. Asymptotically flat scalar, Dirac and Proca stars: discrete vs. continuous families of solutions. *Phys. Lett. B* **2017**, *773*, 654. [CrossRef]
13. Coleman, S.R. Q Balls. *Nucl. Phys. B* **1985**, *262*, 263; Erratum: *Nucl. Phys. B* **1986**, *269*, 744. [CrossRef]
14. Friedberg, R.; Lee, T.D.; Sirlin, A. A Class of Scalar-Field Soliton Solutions in Three Space Dimensions. *Phys. Rev. D* **1976**, *13*, 2739. [CrossRef]
15. Ivanenko, D. Bemerkungen zur theorie der Wechselwirkung. *Sov. Phys.* **1938**, *13*, 41.
16. Weyl, H. A Remark on the Coupling of Gravitation and Electron. *Phys. Rev.* **1950**, *77*, 699. [CrossRef]
17. Heisenberg, W. Doubts and hopes in quantumelectrodynamics. *Physica* **1953**, *19*, 897. [CrossRef]
18. Finkelstein, R.; LeLevier, R.; Ruderman, M. Nonlinear Spinor Fields. *Phys. Rev.* **1951**, *83*, 326. [CrossRef]
19. Finkelstein, R.; Fronsdal, C.F.; Kaus, P. Nonlinear Spinor Field. *Phys. Rev.* **1956**, *103*, 1571. [CrossRef]
20. Soler, M. Classical, stable, nonlinear spinor field with positive rest energy. *Phys. Rev. D* **1970**, *1*, 2766. [CrossRef]
21. Loginov, A.Y. Nontopological solitons in the model of the self-interacting complex vector field. *Phys. Rev. D* **2015**, *91*, 105028. [CrossRef]
22. Friedberg, R.; Lee, T.D.; Pang, Y. Scalar Soliton Stars and Black Holes. *Phys. Rev. D* **1987**, *35*, 3658. [CrossRef]
23. Dzhunushaliev, V.; Folomeev, V. Dirac stars supported by nonlinear spinor fields. *Phys. Rev. D* **2019**, *99*, 084030. [CrossRef]
24. Brihaye, Y.; Delplace, T.; Verbin, Y. Proca Q Balls and their Coupling to Gravity. *Phys. Rev. D* **2017**, *96*, 024057. [CrossRef]
25. Minamitsuji, M. Vector boson star solutions with a quartic order self-interaction. *Phys. Rev. D* **2018**, *97*, 104023. [CrossRef]
26. Deser, S. Absence of Static Solutions in Source-Free Yang-Mills Theory. *Phys. Lett.* **1976**, *64B*, 463. [CrossRef]
27. Volkov, M.S.; Gal'tsov, D.V. Gravitating nonAbelian solitons and black holes with Yang-Mills fields. *Phys. Rept.* **1999**, *319*, 1. [CrossRef]
28. Herdeiro, C.; Radu, E. Construction and physical properties of Kerr black holes with scalar hair. *Class. Quant. Grav.* **2015**, *32*, 144001. [CrossRef]
29. Brito, R.; Cardoso, V.; Herdeiro, C.A.R.; Radu, E. Proca stars: Gravitating Bose–Einstein condensates of massive spin 1 particles. *Phys. Lett. B* **2016**, *752*, 291. [CrossRef]
30. Herdeiro, C.; Radu, E.; Runarsson, H. Kerr black holes with Proca hair. *Class. Quant. Grav.* **2016**, *33*, 154001. [CrossRef]
31. Dolan, S.R.; Dempsey, D. Bound states of the Dirac equation on Kerr spacetime. *Class. Quant. Grav.* **2015**, *32*, 184001. [CrossRef]

32. Brihaye, Y.; Cisterna, A.; Hartmann, B.; Luchini, G. From topological to nontopological solitons: Kinks, domain walls, and *Q*-balls in a scalar field model with a nontrivial vacuum manifold. *Phys. Rev. D* **2015**, *92*, 124061. [CrossRef]

33. Derrick, G.H. Comments on nonlinear wave equations as models for elementary particles. *J. Math. Phys.* **1964**, *5*, 1252–1254. [CrossRef]

34. Gibbons, G.W. Selfgravitating magnetic monopoles, global monopoles and black holes. *Lect. Notes Phys.* **1991**, *383*, 110–138.

35. Kleihaus, B.; Kunz, J.; Radu, E.; Subagyo, B. Axially symmetric static scalar solitons and black holes with scalar hair. *Phys. Lett. B* **2013**, *725*, 489. [CrossRef]

36. Coleman, S.R. The Fate of the False Vacuum. 1. Semiclassical Theory. *Phys. Rev. D* **1977**, *15*, 2929. Erratum: *Phys. Rev. D* 1977, 16, 1248. [CrossRef]

37. Coleman, S.R.; Glaser, V.; Martin, A. Action Minima Among Solutions to a Class of Euclidean Scalar Field Equations. *Commun. Math. Phys.* **1978**, *58*, 211–221. [CrossRef]

38. Heusler, M.; Straumann, N. Scaling arguments for the existence of static, spherically symmetric solutions of self-gravitating systems. *Class. Quant. Grav.* **1992**, *9*, 2177. [CrossRef]

39. Heusler, M. No hair theorems and black holes with hair. *Helv. Phys. Acta* **1996**, *69*, 501.

40. Colpi, M.; Shapiro, S.L.; Wasserman, I. Boson Stars: Gravitational Equilibria of Selfinteracting Scalar Fields. *Phys. Rev. Lett.* **1986**, *57*, 2485. [CrossRef]

41. Pena, I.; Sudarsky, D. Do collapsed boson stars result in new types of black holes? *Class. Quant. Grav.* **1997**, *14*, 3131. [CrossRef]

42. Finster, F.; Yau, J.S.S.T. Nonexistence of time periodic solutions of the Dirac equation in a Reissner-Nordstrom black hole background. *J. Math. Phys.* **2000**, *41*, 2173. [CrossRef]

43. Finster, F.; Smoller, J.; Yau, S.T. Nonexistence of black hole solutions for a spherically symmetric, static Einstein-Dirac-Maxwell system. *Commun. Math. Phys.* **1999**, *205*, 249. [CrossRef]

44. Bachelot-Motet, A. Nonlinear Dirac fields on the Schwarzschild metric. *Class. Quant. Grav.* **1998**, *15*, 1815. [CrossRef]

45. Armendariz-Picon, C.; Greene, P.B. Spinors, inflation, and nonsingular cyclic cosmologies. *Gen. Rel. Grav.* **2003**, *35*, 1637–1658. [CrossRef]

46. Herdeiro, C.; Perapechka, I.; Radu, E.; Shnir, Y. Asymptotically flat spinning scalar, Dirac and Proca stars. *Phys. Lett. B* **2019**, *797*, 134845. [CrossRef]

47. Volkov, M.S.; Wohnert, E. Spinning Q balls. *Phys. Rev. D* **2002**, *66*, 085003. [CrossRef]

48. Kleihaus, B.; Kunz, J.; List, M. Rotating boson stars and Q-balls. *Phys. Rev. D* **2005**, *72*, 064002. [CrossRef]

49. Herdeiro, C.A.R.; Kunz, J.; Perapechka, I.; Radu, E.; Shnir, Y. Multipolar boson stars: Macroscopic Bose-Einstein condensates akin to hydrogen orbitals. *arXiv* **2020**, arXiv:2008.10608.

50. Herdeiro, C.A.R.; Radu, E. Kerr black holes with scalar hair. *Phys. Rev. Lett.* **2014**, *112*, 221101. [CrossRef] [PubMed]

Article

Scalarized Nutty Wormholes

Rustam Ibadov [1], Burkhard Kleihaus [2,*], Jutta Kunz [2] and Sardor Murodov [1]

[1]   Department of Theoretical Physics and Computer Science, Samarkand State University,
      Samarkand 140104, Uzbekistan; ibrustam@mail.ru (R.I.); mursardor@mail.ru (S.M.)
[2]   Institute of Physics, University of Oldenburg, D-26111 Oldenburg, Germany; jutta.kunz@uni-oldenburg.de
*   Correspondence: b.kleihaus@uni-oldenburg.de

Received: 10 December 2020; Accepted: 29 December 2020; Published: 6 January 2021

Abstract: We construct scalarized wormholes with a NUT charge in higher curvature theories. We consider both Einstein-scalar-Gauss-Bonnet and Einstein-scalar-Chern-Simons theories, following Brihaye, Herdeiro and Radu, who recently studied spontaneously scalarised Schwarzschild-NUT solutions. By varying the coupling parameter and the scalar charge we determine the domain of existence of the scalarized nutty wormholes, and their dependence on the NUT charge. In the Gauss-Bonnet case the known set of scalarized wormholes is reached in the limit of vanishing NUT charge. In the Chern-Simons case, however, the limit is peculiar, since with vanishing NUT charge the coupling constant diverges. We focus on scalarized nutty wormholes with a single throat and study their properties. All these scalarized nutty wormholes feature a critical polar angle, beyond which closed timelike curves are present.

Keywords: wormholes; scalar fields; NUT charge; black holes; higher curvature theories

---

1. Introduction

The fascinating phenomenon of scalarization has led to a large variety of interesting observations in the context of compact objects. Scalarization arises, when the generalized Einstein–Klein–Gordon equations lead to solutions with a non-trivial scalar field, caused by the presence of an adequate source term. Depending on the properties of this source term distinct types of scalarized solutions arise. When the source term in the scalar field equation does not vanish for vanishing scalar field, all solutions will be scalarized, and the solutions of ordinary General Relativity (GR) will not be solutions of the coupled set of field equations. In contrast, when the source term in the scalar field equation does vanish for vanishing scalar field the GR solutions do remain solutions of the generalized set of field equations. However, they develop tachyonic instabilities, where new scalarized solutions arise.

The latter phenomenon was first observed for neutron stars in scalar-tensor theories [1] where it is referred to as matter-induced spontaneous scalarization, since the source term for the scalar field is provided by the highly compact nuclear matter. Only much more recently it was observed that in the case of the vacuum black holes of GR spontaneous scalarization is possible as well, when another type of source term for the scalar field is provided. Coupling the higher curvature Gauss–Bonnet (GB) invariant to the scalar field with an appropriate coupling function curvature-induced spontaneously scalarized black holes arise, representing scalarized Schwarzschild and Kerr black holes [2–24]. Moreover, the curvature-induced spontaneous scalarized Kerr black holes come in two types, those that arise in the slow rotation limit from the scalarized Schwarzschild black holes, and those that do not possess a slow rotation limit [25–27]. We note that for GB theories with coupling functions allowing for spontaneous scalarization the current constraints from gravitational waves produced in binary black hole mergers can be avoided, since the scalar field can be set to zero in cosmological applications, implying the cosmological standard results. In the case of localized compact objects, however, the deviation of the speed of gravitational waves from the speed of light decays rapidly with distance [28–30].

An alternative higher curvature invariant to study curvature-induced spontaneously scalarized black holes is the Chern–Simons (CS) invariant. However, in the static case of the Schwarzschild metric the invariant vanishes, and therefore no spontaneously scalarized Schwarzschild black holes arise. This is different for the Kerr metric, since rotation leads to a finite CS source term for the scalar field [31–38], which should allow for spontaneously scalarized Kerr black holes. In order to learn about scalarized rotating CS black holes without having to deal with the full complexity of the challenging set of the resulting coupled partial differential equations, one may first resort to the technically much simpler case and include a NUT charge [39–45] instead of rotation, as pursued successfully by Brihaye et al. [46,47]. In that case, a much simpler set of ordinary differential equations (ODEs) results, since the angular dependence of the scalarized solutions factorizes.

Inspired by Brihaye et al. [46,47], we here follow their motivation and apply this strategy to wormholes, constructing scalarized nutty wormholes in higher curvature theories employing either a GB term or a CS term as source term for the scalar field. Following Brihaye et al. [46], we here employ a quadratic coupling function. The spontaneously scalarized Schwarzschild-NUT solutions of [46] then represent one of the boundaries of the domain of existence of the scalarized nutty wormholes. However, the wormhole solutions are not spontaneously scalarized, since when the scalar field vanishes, pure vacuum GR is retained, and there are no Lorentzian traversable wormhole solutions in vacuum GR (see e.g., [48–52]).

To obtain traversable wormhole solutions the energy conditions must be violated. In GR this can be achieved by the presence of exotic matter. However, by allowing for alternative theories of gravity traversable wormholes can be obtained without the need for exotic matter (see e.g., [53–63]). Employing the string theory motivated dilaton-GB coupling, static scalarized wormholes were shown to exist, and their domain of existence and their properties were studied before [59–61]. In the dilatonic case, the black hole boundary of the domain of existence corresponds to static dilatonic black holes [64]. For other coupling functions, which give rise to spontaneously scalarized GB black holes, the black hole boundary of the domain of existence of scalarized wormholes [59,65] consists of the corresponding spontaneously scalarized black holes [2–4].

The domain of existence of scalarized wormholes is further bounded by a set of solutions, where singularities are encountered, and by a set of solutions, where the wormhole throat becomes degenerate [59–61,65]. In the latter case, this degeneracy reveals, that in addition to wormholes with a single throat also wormholes with an equator and a double throat exist. The throat(s) and equator arise naturally in these solutions in one of the two parts of the spacetime. However, when simply continuing the solutions beyond the throat (or equator) a singularity will invariably be encountered. In order to obtain wormholes without such singularities, symmetry has been imposed with respect to the throat (or equator). This entails that a thin shell of matter is needed at the throat (or equator), to satisfy the respective Israel junction conditions [66,67]. Typically, this matter can be ordinary matter, thus no exotic matter is needed [59–61,65].

Here we generalize these scalarized wormhole solutions in two ways. On the one hand, we include a NUT charge and on the other hand we consider besides the GB invariant also the CS invariant. The presence of the NUT charge implicates a Misner string on the polar axis. Therefore the resulting spacetimes are not asymptotically flat in the usual sense. However, all unknown functions of the wormhole solutions are only functions of the radial coordinate, and their asymptotic fall-off is of the usual type of an asymptotically flat spacetime, as in the case of scalarized Schwarzschild-NUT solutions [46,47]. As always, the presence of a NUT charge gives rise to closed timelike curves, since the metric component of the azimuthal angle, $g_{\varphi\varphi}$, changes sign in the manifold at a critical value of the polar angle θ. Here we show that all scalarized nutty wormhole solutions possess such a critical polar angle at their throat. In the black hole limit the throat changes into a horizon, where the critical polar angle goes to zero.

The paper is organized as follows: In Section 2 we present the actions involving the GB and the CS term and exhibit the equations of motion for both cases. We then discuss the boundary conditions, the conditions for the center, i.e., the throat (or equator), the junction conditions, and the null energy condition (NEC). Subsequently we address the numerical procedure and present our results in Section 3. These include, in particular, the profile functions for the scalarized nutty wormhole solutions, the violation of the NEC, the domain of existence with its outer boundaries, and an analysis of the junction conditions for the thin shell of matter at the throat. We give our conclusions and an outlook in Section 4.

2. Theoretical Setting

2.1. Action and Equations of Motion

Following Brihaye et al. [46], we consider the effective action for Einstein-scalar-higher curvature invariant theories

$$S = \frac{1}{16\pi} \int \left[R - \frac{1}{2} \partial_\mu \phi \, \partial^\mu \phi + F(\phi) \mathcal{I}(g) \right] \sqrt{-g} d^4 x \, , \tag{1}$$

where R is the curvature scalar, and ϕ denotes the massless scalar field without self-interaction, that is coupled with some coupling function $F(\phi)$ to an invariant $\mathcal{I}(g)$. For the coupling function $F(\phi)$ we choose a quadratic ϕ-dependence with coupling constant α,

$$F(\phi) = \alpha \phi^2 \, , \tag{2}$$

the simplest choice leading to curvature-induced spontaneous scalarization of black holes.

For the invariant $\mathcal{I}(g)$ we make two choices, (i) the Gauss–Bonnet term

$$\mathcal{I}(g) = R^2_{\text{GB}} = R_{\mu\nu\rho\sigma} R^{\mu\nu\rho\sigma} - 4 R_{\mu\nu} R^{\mu\nu} + R^2 \, , \tag{3}$$

and (ii) the Chern–Simons term

$$\mathcal{I}(g) = R^2_{\text{CS}} = {}^* R^\mu{}_\nu{}^{\rho\sigma} R^\nu{}_{\mu\rho\sigma} \, , \tag{4}$$

where the Hodge dual of the Riemann-tensor ${}^* R^\mu{}_\nu{}^{\rho\sigma} = \frac{1}{2} \eta^{\rho\sigma\kappa\lambda} R^\mu{}_{\nu\kappa\lambda}$ is defined with the 4-dimensional Levi-Civita tensor $\eta^{\rho\sigma\kappa\lambda} = \epsilon^{\rho\gamma\sigma\tau} / \sqrt{-g}$. While both invariants are topological in four dimensions, the coupling to the scalar field ϕ via the coupling function $F(\phi)$ provides significant contributions to the equations of motion.

We obtain the coupled set of field equations by varying the action (1) with respect to the scalar field and to the metric,

$$\nabla^\mu \nabla_\mu \phi + \frac{dF(\phi)}{d\phi} \mathcal{I} = 0 \, , \tag{5}$$

$$G_{\mu\nu} = \frac{1}{2} T^{(\text{eff})}_{\mu\nu} \, , \tag{6}$$

where $G_{\mu\nu}$ is the Einstein tensor and $T^{(\text{eff})}_{\mu\nu}$ denotes the effective stress-energy tensor

$$T^{(\text{eff})}_{\mu\nu} = T^{(\phi)}_{\mu\nu} + T^{(\mathcal{I})}_{\mu\nu} \, , \tag{7}$$

which consists of the scalar field contribution

$$T^{(\phi)}_{\mu\nu} = (\nabla_\mu \phi)(\nabla_\nu \phi) - \frac{1}{2} g_{\mu\nu} (\nabla_\rho \phi)(\nabla^\rho \phi) \, , \tag{8}$$

and a contribution from the respective invariant $\mathcal{I}(g)$. For the chosen invariants we obtain (i)

$$T_{\mu\nu}^{(GB)} = \left(g_{\rho\mu}g_{\lambda\nu} + g_{\lambda\mu}g_{\rho\nu}\right)\eta^{\kappa\lambda\alpha\beta}{}^*R^{\rho\gamma}{}_{\alpha\beta}\nabla_\gamma\nabla_\kappa F(\phi) , \tag{9}$$

and (ii)

$$T^{(CS)\mu\nu} = -8\left[\nabla_\rho F(\phi)\right]\epsilon^{\rho\sigma\tau(\mu}\nabla_\tau R^{\nu)}{}_\sigma + \left[\nabla_\rho\nabla_\sigma F(\phi)\right]{}^*R^{\sigma(\mu\nu)\rho}. \tag{10}$$

To obtain static, spherically symmetric wormhole solutions with a NUT charge N we assume the line element to be of the form

$$ds^2 = -e^{f_0}\left(dt - 2N\cos\theta d\varphi\right)^2 + e^{f_1}\left[dr^2 + r^2\left(d\theta^2 + \sin^2\theta d\varphi^2\right)\right] , \tag{11}$$

All three functions, the two metric functions f_0 and f_1 and the scalar field function ϕ, depend only on the radial coordinate r.

When we insert the above ansatz (11) for the metric and the scalar field into the scalar-field Equation (5) and the Einstein Equation (6) with effective stress-energy tensor (7), we obtain five coupled, nonlinear ODEs. However, these are not independent, since the θ-dependence factorizes, and one ODE can be treated as a constraint. This leaves us with three coupled ODEs of second order. Note that in case (i) the system can be reduced to one first order and two second order ODEs.

Inspection of the field equations reveals an invariance under the scaling transformation

$$r \to \chi r , \quad N \to \chi N , \quad t \to \chi t , \quad F \to \chi^2 F , \tag{12}$$

with constant $\chi > 0$.

2.2. Throats, Equators, and Boundary Conditions

In order to obtain scalarized nutty wormhole solutions, we need to impose an appropriate set of boundary conditions for the ODEs, which we now address. We first introduce the circumferential (or spherical) radius

$$R_C = e^{\frac{f_1}{2}} r \tag{13}$$

of the wormhole spacetimes, which may possess one or more finite extrema. If there is a single finite extremum, this corresponds to the single throat of the respective wormholes. Here we will mainly consider such single throat wormholes, thus featuring a single minimum. But wormholes with more extrema may also exist. They might, for instance, possess a local maximum surrounded by two minima. The local maximum would then correspond to their equator, while the two minima would represent their two throats, making them double throat wormholes.

To obtain the first set of boundary conditions we therefore require the presence of an extremum of the spherical radius at some $r = r_0$. This yields

$$\left.\frac{dR_C}{dr}\right|_{r=r_0} = 0 \iff \left.\frac{df_1}{dr}\right|_{r=r_0} = -\frac{2}{r_0} . \tag{14}$$

Some details on the condition $R_C''(r_0) > 0$ are given in the Appendix A. In the following we will refer to the two-dimensional submanifolds defined by $r = r_0$ and $t = \text{const.}$ as the center of the configurations.

We note that the presence of a NUT charge leads to an interesting feature of these wormholes. Unlike the usual case, the throat metric of these nutty wormholes

$$ds_{\text{th}}^2 = e^{f_1(r_0)}r_0^2\left(d\theta^2 + \left[\sin^2\theta - \frac{4N^2}{r_0^2}e^{f_0(r_0)-f_1(r_0)}\cos^2\theta\right]d\varphi^2\right) \tag{15}$$

changes its signature at the critical angles θ_c and $\pi - \theta_c$, where the coefficient of $d\varphi^2$ changes sign. We obtain θ_c from the condition $\det(g_{\text{th}}) \geq 0$, which requires $\theta_c \leq \theta \leq \pi - \theta_c$ with

$$\theta_c = \arctan\left(\frac{2|N|}{r_0} e^{\frac{f_0 - f_1}{2}}\right)\Bigg|_{r_0} . \tag{16}$$

Only for $\theta_c \leq \theta \leq \pi - \theta_c$ the signature of the metric is positive, as required for a two-dimensional Riemannian surface. The change of signature is a consequence of the non-causal structure of a spacetime in the presence of a NUT charge N, which allows for closed timelike curves.

We obtain the second set of boundary conditions by requiring the usual boundary conditions for $r \to \infty$ [46]. The associated asymptotic expansions of the metric functions and the scalar field read

$$f_0 = -\frac{2M}{r} + \mathcal{O}\left(r^{-3}\right), \tag{17}$$

$$f_1 = \frac{2M}{r} + \mathcal{O}\left(r^{-2}\right), \tag{18}$$

$$\phi = \phi_\infty - \frac{D}{r} + \mathcal{O}\left(r^{-3}\right), \tag{19}$$

where M denotes the mass of the wormholes and D corresponds to their scalar charge. The quantity ϕ_∞ represents the asymptotic value of the scalar field. We note that all higher order terms in the expansion can be expressed in terms of M, D and ϕ_∞. Thus the solution is uniquely determined by these quantities (and parameters of the theory). Since we are interested in the relation of the wormhole solutions to the spontaneously scalarized black hole solutions of Brihaye et al. [46] we need to choose the same asymptotic value $\phi_\infty = 0$.

2.3. Junction Conditions

In order to obtain wormholes whose geometry is symmetric with respect to the center and which do not possess any singularity (apart from the Misner string), we need to impose junction conditions at the center. For the discussion of the junction conditions it is useful to introduce the radial coordinate η,

$$\eta = r_0\left(\frac{r}{r_0} - \frac{r_0}{r}\right), \tag{20}$$

where r_0 is a constant. We then define the constant η_0 via $\eta_0 = 2r_0$. In terms of the new radial coordinate η the metric reads

$$ds^2 = -e^{f_0}\left(dt - 2N\cos\theta d\varphi\right)^2 + e^{F_1}\left[d\eta^2 + \left(\eta^2 + \eta_0^2\right)\left(d\theta^2 + \sin^2\theta d\varphi^2\right)\right], \tag{21}$$

where we have introduced the new metric function F_1,

$$e^{F_1} = e^{f_1}\left(1 + \frac{r_0^2}{r^2}\right)^{-2} .$$

Thus the center is located at $\eta = 0$, and the solution on the $\eta \leq 0$ part of the manifold can be obtained from the solution on the $\eta \geq 0$ part of the manifold by imposing the symmetry conditions $f_0(-\eta) = f_0(\eta)$, $F_1(-\eta) = F_1(\eta)$ and $\phi(-\eta) = \phi(\eta)$. However, these conditions generically introduce jumps in the derivatives of the functions f_0 and ϕ at the center $\eta = 0$, which may be attributed to a thin shell of matter that is localized at the center.

To properly embed such a thin shell of matter in the complete wormhole solution, we make use of an appropriate set of junction conditions [66,67]. In particular, we consider jumps in the coupled set of Einstein and scalar field equations that arise when $\eta \to -\eta$,

$$\langle G^{\mu}{}_{\nu} - T^{\mu}{}_{\nu} \rangle = s^{\mu}{}_{\nu}\,, \quad \langle \nabla^2 \phi + \dot{F}\mathcal{I} \rangle = s_{\text{scal}}\,, \tag{22}$$

with the abbreviation $dF(\phi)/d\phi = \dot{F}$ for the derivative of the coupling function. Here we have denoted the stress-energy tensor of the matter at the center by $s^{\mu}{}_{\nu}$, and the source term for the scalar field by s_{scal}. We would like the matter forming the thin shells to be some form of ordinary (non-exotic) matter. We will therefore assume that there is a perfect fluid at the center which has pressure p and energy density ϵ_c, and that there is a scalar charge density ρ_{scal} together with a gravitational source [60,61]

$$S_{\Sigma} = \int [\lambda_1 + 2\lambda_0 F(\phi)\bar{R}]\,\sqrt{-\bar{h}}d^3x\,, \tag{23}$$

that has been employed before for GB wormholes without NUT charge. Here we have introduced the constants λ_1 and λ_0, $\bar{h}_{ab}$ denotes the induced metric at the center, and $\bar{R}$ denotes the associated Ricci scalar. In order to obtain the junction conditions, we substitute the metric into the set of equations (22).

Now we derive the junction conditions for both invariants separately. Note, that here and in the following all functions and derivatives are evaluated at the center. In the case of the Gauss–Bonnet invariant we find the equations

$$\frac{4}{\eta_0^2}\dot{F}\phi'\left(\eta_0^2 e^{-\frac{3}{2}F_1} + 3N^2 e^{f_0 - \frac{5}{2}F_1}\right) = \lambda_1\eta_0^2 + 4\lambda_0 F \frac{\eta_0^2 e^{-F_1} + 3N^2 e^{f_0 - 2F_1}}{\eta_0^2} - \epsilon_c \eta_0^2\,, \tag{24}$$

$$N\cos\theta\left[\eta_0^2 f_0' e^{-\frac{F_1}{2}} - 8\dot{F}\phi'\left(e^{-\frac{3}{2}F_1} + \frac{4N^2}{\eta_0^2} e^{f_0 - \frac{5}{2}F_1}\right)\right] = 2N\cos\theta\left[(\epsilon_c + p)\eta_0^2 - 4\lambda_0 F \frac{\eta_0^2 e^{-F_1} + 4N^2 e^{f_0 - 2F_1}}{\eta_0^2}\right]\,, \tag{25}$$

$$\frac{\eta_0^2 f_0'}{2} e^{-\frac{F_1}{2}} - \frac{4N^2}{\eta_0^2}\dot{F}\phi' e^{f_0 - \frac{5}{2}F_1} = p\eta_0^2 + \lambda_1\eta_0^2 - 4\lambda_0 N^2 F \frac{e^{f_0 - 2F_1}}{\eta_0^2}\,, \tag{26}$$

$$e^{-F_1}\phi' - 4\frac{\dot{F}}{\eta_0^4} f_0'\left(\eta_0^2 e^{-2F_1} + 3N^2 e^{f_0 - 3F_1}\right) = -4\lambda_0\frac{\dot{F}}{\eta_0^4}\left(\eta_0^2 e^{-F_1} + N^2 e^{f_0 - 2F_1}\right) + \frac{\rho_{\text{scal}}}{2}\,, \tag{27}$$

which follow from the $\left({}^t{}_t\right)$, $\left({}^t{}_\varphi\right)$, and $\left({}^\varphi{}_\varphi\right)$ components of the Einstein equations and from the scalar field equation, respectively. Note that the $\left({}^\theta{}_\theta\right)$ equation is equivalent to the $\left({}^\varphi{}_\varphi\right)$ equation, and that all other equations are satisfied trivially. We also remark that the θ dependence in the $\left({}^t{}_\varphi\right)$ equation factorizes, and that this equation is satisfied once the $\left({}^t{}_t\right)$ and $\left({}^\varphi{}_\varphi\right)$ equations are solved.

As an example we consider pressureless matter, $p = 0$. With

$$\lambda_0 = \frac{\dot{F}}{F} e^{-\frac{F_1}{2}}\phi'\,, \quad \lambda_1 = \frac{f_0'}{2} e^{-\frac{F_1}{2}}\,, \tag{28}$$

we find the simple result

$$\epsilon_c = \frac{f_0'}{2} e^{-\frac{F_1}{2}}\,. \tag{29}$$

Since for our solutions $f_0' > 0$ the energy density ϵ_c is always positive for this choice of the constants λ_0 and λ_1.

We now turn to case of the Chern–Simons invariant where we find the equations

$$8N\dot{F}\phi' f_0' e^{\frac{f_0}{2}} e^{-2F_1} = \lambda_1 \eta_0^2 + 4\lambda_0 F \frac{\eta_0^2 e^{-F_1} + 3N^2 e^{f_0 - 2F_1}}{\eta_0^2} - \epsilon_c \eta_0^2 , \tag{30}$$

$$N \cos\theta f_0' \left(\eta_0^2 e^{-\frac{F_1}{2}} - 24N\dot{F}\phi' e^{\frac{f_0}{2} - 2F_1} \right) = 2N \cos\theta \left(\epsilon_c + p\right) \eta_0^2 - 8N \cos\theta \lambda_0 F \frac{\eta_0^2 e^{-F_1} + 4N^2 e^{f_0 - 2F_1}}{\eta_0^2} , \tag{31}$$

$$\frac{f_0'}{2} \left(\eta_0^2 e^{-\frac{F_1}{2}} - 8N\dot{F}\phi' e^{\frac{f_0}{2} - 2F_1} \right) = p\eta_0^2 + \lambda_1 \eta_0^2 - 4\lambda_0 N^2 F \frac{e^{f_0 - 2F_1}}{\eta_0^2} , \tag{32}$$

$$e^{-F_1} \phi' - 4N \frac{\dot{F}}{\eta_0^2} (f_0')^2 e^{\frac{f_0 - 5F_1}{2}} = -4\lambda_0 \frac{\dot{F}}{\eta_0^4} \left(\eta_0^2 e^{-F_1} + N^2 e^{f_0 - 2F_1} \right) + \frac{p_{\text{scal}}}{2} , \tag{33}$$

which follow again from the $\left(^t{}_t\right)$, $\left(^t{}_\varphi\right)$, $\left(^\varphi{}_\varphi\right)$ components of the Einstein equations and the scalar field equation, respectively.

Again we consider as an example pressureless matter, $p = 0$. With

$$\lambda_0 = \frac{\eta_0^2 \dot{F}}{NF} e^{-\frac{f_0}{2}} , \quad \lambda_1 = -\frac{f_0'}{2} e^{\frac{F_1}{2}} , \tag{34}$$

we now find

$$\epsilon_c = \frac{f_0'}{2N} \left(Ne^{-\frac{F_1}{2}} + 8\dot{F}\phi' e^{-\frac{f_0}{2} - F_1} \right) + \frac{4N}{\eta_0^2} \dot{F}\phi' f_0' e^{\frac{f_0}{2} - 2F_1} . \tag{35}$$

2.4. Energy Conditions

In wormhole solutions the null energy condition (NEC)

$$T_{\mu\nu} n^\mu n^\nu \geq 0 \tag{36}$$

must be violated, where n^μ is any null vector ($n^\mu n_\mu = 0$). Thus it is sufficient to show that null vectors exist, such that $T_{\mu\nu} n^\mu n^\nu < 0$ in some spacetime region. Such a null vector n^μ is given by

$$n^\mu = \left(1, \sqrt{-g_{tt}/g_{\eta\eta}}, 0, 0 \right) , \tag{37}$$

and thus $n_\mu = \left(g_{tt}, \sqrt{-g_{tt}\, g_{\eta\eta}}, 0, 0 \right)$. The NEC then takes the form

$$T_{\mu\nu} n^\mu n^\nu = T_t^t n^t n_t + T_\eta^\eta n^\eta n_\eta = -g_{tt} \left(-T_t^t + T_\eta^\eta \right) . \tag{38}$$

Consequently the NEC is violated when

$$- T_t^t + T_\eta^\eta < 0 . \tag{39}$$

Alternatively, considering the null vector

$$n^\mu = \left(1, 0, \sqrt{-g_{tt}/g_{\theta\theta}}, 0 \right) , \tag{40}$$

the NEC is violated when

$$- T_t^t + T_\theta^\theta < 0 . \tag{41}$$

These conditions have been addressed before for various scalarized wormhole solutions [48,59–61]. We analyze these conditions in Section 3.2 for the scalarized nutty wormholes.

3. Results

3.1. Numerics

In order to solve the coupled Einstein and scalar field equations numerically we introduce the inverse radial coordinate $x = 1/r$. The asymptotic region $r \to \infty$ then corresponds to $x \to 0$. In this region the expansion of the metric functions and the scalar field reads (see Equations (17)–(19))

$$f_0 = -2Mx + \mathcal{O}\left(x^3\right), \quad f_1 = 2Mx + \mathcal{O}\left(x^2\right), \quad \phi = \phi_\infty - Dx + \mathcal{O}\left(x^3\right). \tag{42}$$

We treat the system of ODEs as an initial value problem, for which we employ the fourth order Runge Kutta method. From the above expansion we read off the initial values,

$$f_{0,\mathrm{ini}} = 0, \quad f'_{0,\mathrm{ini}} = -2M, \quad f_{1,\mathrm{ini}} = 0, \quad f'_{1,\mathrm{ini}} = 2M, \quad \phi_{\mathrm{ini}} = 0, \quad \phi'_{\mathrm{ini}} = -D. \tag{43}$$

The computation then starts at spatial infinity, $x = 0$, and ends at the center at some finite $x = x_0$, where the condition (14) is reached. Note that the initial conditions determine the solution uniquely, as discussed in Section 2.2 There is no fine-tuning of parameters needed.

3.2. Solutions

By following the above numerical procedure we obtain the sets of nutty wormhole solutions for both invariants $\mathcal{I}(g)$. Here we demonstrate some typical solutions for both cases. We exhibit in Figure 1 the metric profile functions e^{f_0}, e^{F_1}, and the scalar field function ϕ versus the radial coordinate η/M for the GB invariant (a) and the CS invariant (b), choosing parameters $\alpha/M^2 = 2.5$, $D/M = 2$ and $n = N/M = 3$, and $\alpha/M^2 = 4$, $D/M = 1$ and $n = N/M = 3$, respectively. The figures also show the circumferential radius R_c/η_0 versus η/M ($\eta_0/M = 0.517$: GB invariant, $\eta_0/M = 0.495$: CS invariant). As required, R_c reaches an extremum at the center $\eta = 0$. We note that the solutions have rather similar properties for both invariants.

To see that the wormhole solutions violate the energy conditions, we inspect the components of the effective energy momentum tensor, T_t^t, T_ϕ^t, T_η^η, and T_θ^θ. These are shown for the same solutions and the GB and CS invariants in Figure 1c,d, respectively. In particular, we note, that the component T_η^η is negative in the vicinity of the center for both invariants. Moreover, all components are negative in some region of the spacetime. We exhibit in Figure 1e,f the NEC conditions $-T_t^t + T_\eta^\eta \geq 0$ and $-T_t^t + T_\theta^\theta \geq 0$ for the GB and CS invariants, respectively. The figures clearly demonstrate the NEC violation for both invariants.

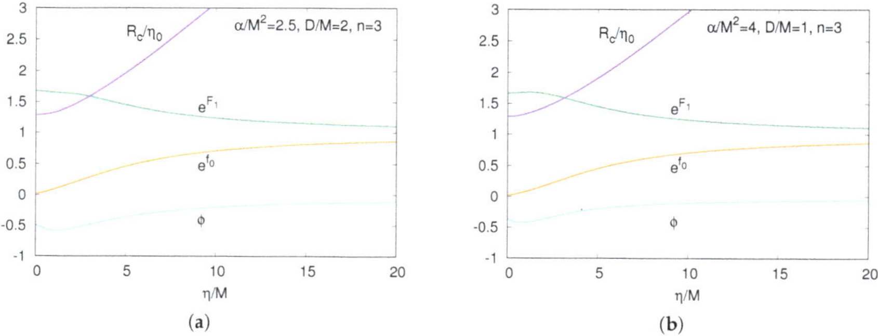

Figure 1. *Cont.*

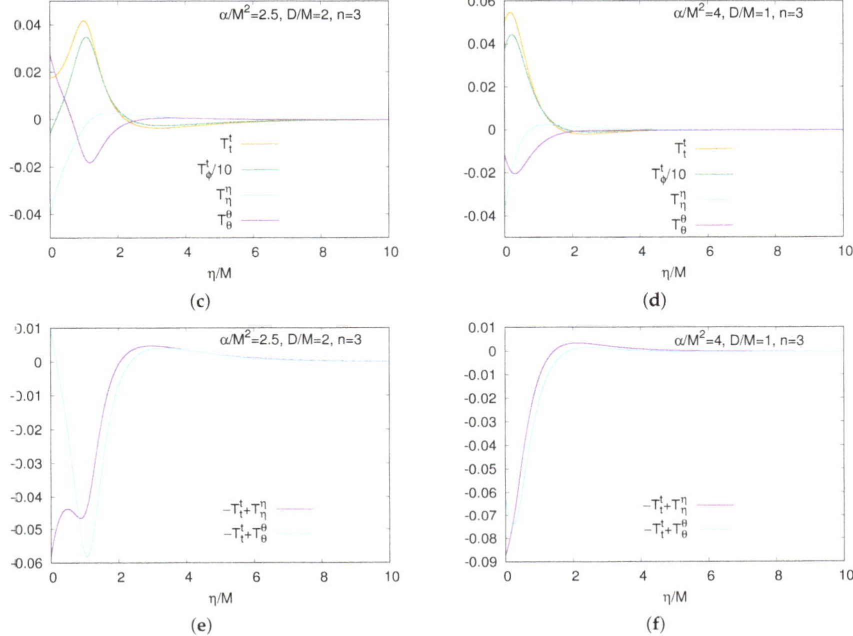

Figure 1. Examples of nutty wormhole solutions (left plots: Gauss–Bonnet with parameters $\alpha/M^2 = 2.5$, $D/M = 2$ and $n = N/M = 3$, right plots: Chern–Simons with $\alpha/M^2 = 4$, $D/M = 1$ and $n = N/M = 3$): (**a,b**) metric profile functions e^{f_0}, e^{F_1}, scalar field function ϕ, and scaled circumferential radius R_c/η_0 vs radial coordinate η; (**c,d**) stress-energy tensor components T^t_t, T^t_ϕ, T^η_η, and T^θ_θ vs. radial coordinate η/M; (**e,f**) NEC conditions $-T^t_t + T^\eta_\eta \geq 0$ and $-T^t_t + T^\theta_\theta \geq 0$ vs radial coordinate η.

3.3. Domain of Existence

We now address the domain of existence of these nutty wormhole solutions. We exhibit the domain of existence in Figure 2 for the GB (left) and CS (right) invariants, for a set of values of the scaled NUT charge $n = N/M$. In particular, we show the outer boundaries of the respective domains of existence, by presenting the scaled coupling constant α/M^2 versus the scaled scalar charge D/M.

The existence of the wormhole solutions requires the presence of a non-trivial scalar field. Since the equations are invariant under the transformation $\phi \to -\phi$, the domain of existence is symmetric with respect to $D \to -D$, and it is sufficient to only exhibit $D \geq 0$. For $D = 0$ the scalar field vanishes, and thus $D = 0$ represents the trivial boundary of the domain of existence, where pure GR solutions reside. The first non-trivial boundary corresponds to scalarized nutty black holes, and is shown by the solid red curves. These scalarized black holes were obtained before by Brihaye et al. [46]. Here the center/throat turns into a black hole horizon.

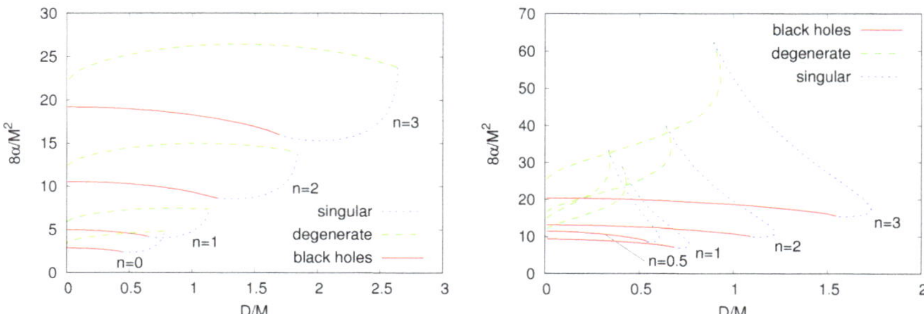

Figure 2. Domain of existence ((**left**) plot: Gauss–Bonnet, (**right**) plot: Chern–Simons) for several values of the scaled NUT charge $n = N/M$: scaled coupling constant α/M^2 vs scaled scalar charge D/M. The solid red curves represent the black hole limit, the dashed green curves the degenerate wormhole limit, and the dotted blue curves the singular limit.

The second non-trivial boundary in the figures is termed degenerate and shown by the dashed green curves. To understand this boundary, we recall the numerical construction of the solutions. The calculation is ended, when an extremum is reached. However, in principle, we can continue the calculation beyond the extremum, where we might find a second extremum. The first one then corresponds to a throat while the second one corresponds to an equator. As the coupling constant is varied, the two extrema will approach each other until finally a degenerate extremum is reached. This third boundary represents precisely the values of the coupling constant, where such a degenerate extremum is reached.

The last boundary has been labeled singular, and is shown by the dotted blue curves. At this boundary the calculations reveal the appearance of a singularity somewhere in the spacetime, that is of the cusp type (see [59,65,68,69]). Their presence is due to the emergence of a node at some value of the radial coordinate $\eta_\star$ in the determinant, that arises upon diagonalisation of the ODEs. These cusp singularities seem to be a rather common feature of scalarized wormholes. Here we see, that they do not only arise for GB theories, but are also present for CS theories.

The figures also show, that the domain of existence of wormhole solutions increases strongly with increasing NUT charge. For the GB coupling, the limit of vanishing NUT charge leads to the scalarized wormhole solutions of Antoniou et al. [59]. For the CS coupling, the vanishing of the NUT charge needs special attention, since in this case two branches of solutions arise, as shown for the Schwarzschild-NUT solutions by Brihaye et al. [46]. Along the first branch the coupling constant decreases with decreasing NUT charge, analogous to the GB case. However, along the second branch the coupling constant increases with decreasing NUT charge. This is seen in Figure 2(right), where the solutions for $n = N/M = 3, 2$ and 1 are on the first branch, while the solutions for $n = N/M = 0.5$ are already on the second branch. Interestingly, the smaller the values of the NUT charge, the larger the values of the coupling constant that are necessary to obtain scalarized solutions. In fact, both conspire in such a way, that a finite domain arises also in the CS case in the limit $N \to 0$ as further discussed below.

3.4. Throat Properties

We next address properties of the wormhole center, and since we do not consider equators here, but only throats, we refer to these properties as throat properties. In particular, we consider the value of the circumferential radius R_C, Equation (13), at the throat which we denote by R_{th}, the value of the metric function e^{f_0}, and the value of the critical angle θ_c, Equation (16). For these properties we delimit their domains of existence in Figure 3, by the extracting the boundaries of the respective domains. As before we show the black hole limit by solid red curves, the degenerate limit by dashed green curves, and the singular limit by dotted blue curves.

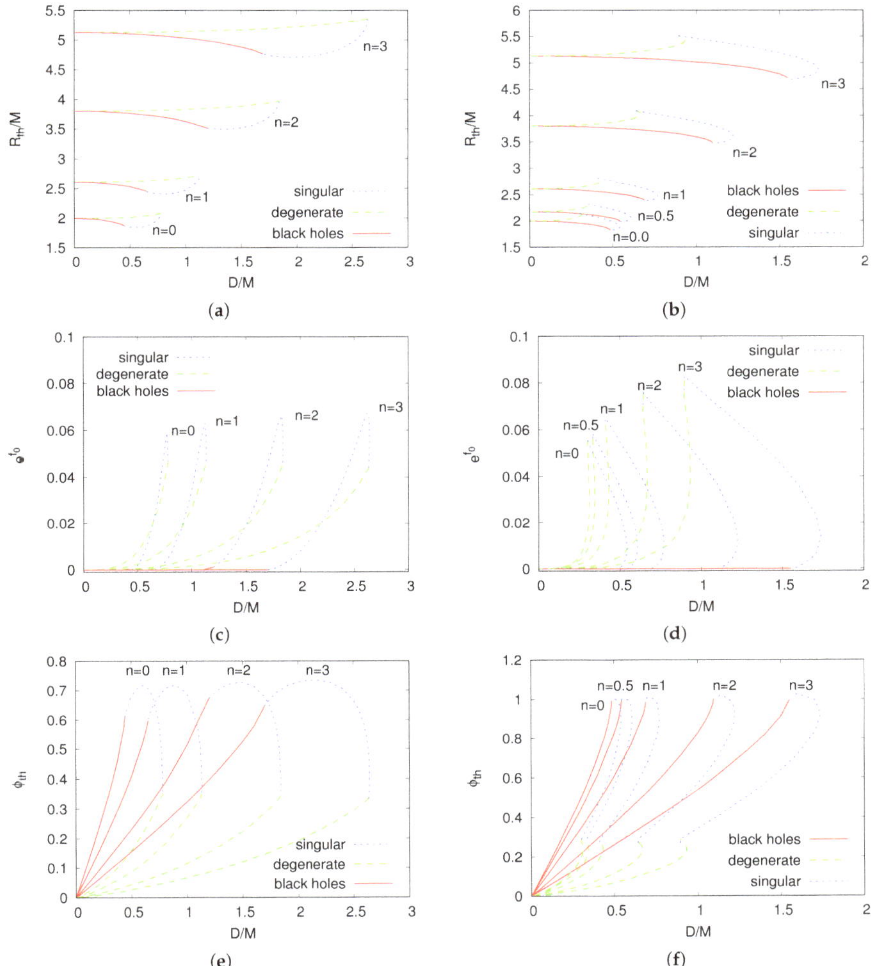

Figure 3. Properties at the throat (left plots: Gauss–Bonnet, right plots: Chern–Simons) for several values of the scaled NUT charge $n = N/M$: (**a**,**b**) scaled circumferential radius R_{th}/M, (**c**,**d**) metric function e^{f_0}, (**e**,**f**) scalar field ϕ_{th} vs scaled scalar charge D/M. The solid red curves represent the black hole limit, the dashed green curves the degenerate wormhole limit, and the dotted blue curves the singular limit.

We exhibit the domain of the scaled value R_{th}/M versus the scaled scalar charge D/M in Figure 3 for the GB invariant (a) and the CS invariant (b) for several values of the NUT charge $n = N/M$. We note that for a fixed value of the NUT charge $n = N/M$, in the limit of vanishing scalar charge ($D/M \rightarrow 0$) the same solution is approached for both the GB invariant (a) and the CS invariant (b). In this limit the scalar field vanishes identically, and thus there are no wormhole solutions. The limiting solution therefore corresponds to a Schwarzschild-NUT black hole solution, for which the value of R_{th}/M depends only on the value of the NUT charge $n = N/M$.

When taking the NUT charge to zero, the boundary composed of scalarized black holes remains finite. In the GB case the scalarized Schwarzschild solutions are approached. In the CS case the coupling constant diverges, as the NUT charge goes to zero, leaving a finite value for the source term of the scalar field. In fact a new coupling constant might be considered, $\alpha' = \alpha/N$ such that α' is finite and might be varied. Thus a peculiar set of limiting solutions arises, that starts from the Schwarzschild black hole with $R/M = 2$ at $D/M = 0$ and forms the black hole boundary.

Considering the full domain of existence, in the GB case in the limit $n = N/M \to 0$ the known finite domain of scalarized wormhole solutions is approached [59]. In the CS case the limit $n = N/M \to 0$ leads to a finite domain of solutions as well, albeit solutions resulting from a peculiar cancellation. Nevertheless, the change of the domain of existence of the circumferential radius R_{th}/M is completely smooth in the limit $n = N/M \to 0$, as calculations for several small values of $n = N/M$ have shown. The limiting domain is seen in Figure 3b. In this figure it does not make a difference whether the solutions are on the first (large $n = N/M$) or second (small $n = N/M$) branch.

Inspection of the metric function e^{f_0}, shown versus the scaled scalar charge D/M in Figure 3 for the GB invariant (c) and the CS invariant (d), yields full agreement with the above discussion. In the black hole limit the metric function vanishes as it must, while the boundary of degenerate scalarized wormholes approaches also the black hole limit in the limit of vanishing scalar charge, $D/M \to 0$. On the other hand, the boundary of singular wormholes connects to the scalarized nutty black holes and contains the maximal value of this metric function for a given value of the NUT charge $n = N/M$. This maximum remains finite for $n = N/M \to 0$.

The scalar field at the throat ϕ_{th} is exhibited versus the scaled scalar charge D/M in Figure 3 for the GB invariant (e) and the CS invariant (f). As argued above, for $D/M \to 0$ the scalar field vanishes, and thus also $\phi_{\mathrm{th}} \to 0$. Analogous to the metric function e^{f_0}, for a given NUT charge $n = N/M$ a maximum of ϕ_{th} is reached on the singular boundary, and for $n = N/M \to 0$ the maximum remains finite.

3.5. Junction Conditions and Critical Polar Angle

As discussed above we need to satisfy junction conditions in order to obtain symmetric wormholes without singularities (except for the Misner string). We have therefore introduced an action at the center and allowed for a thin shell of matter in the form of a perfect fluid with energy density ϵ_c. Since we here focus only on wormholes with a single throat, we denote this energy density by ϵ_{th}. We exhibit in Figure 4 ϵ_{th} for a pressureless fluid for several sets of wormhole solutions with fixed values of the scaled scalar charge D/M and NUT charge $n = N/M = 1$ versus the scaled coupling constant α/M^2 for the GB invariant (a) and the CS invariant (b).

The boundaries of these sets of solutions are given by the dashed green curves, representing the degenerate wormhole limit, and the dotted blue curves, representing the singular limit. We note, that the energy density ϵ_{th} is positive in all cases shown. Thus these wormholes need only ordinary matter at the throat to remain open.

Last we address the critical polar angle at the center θ_c, Equation (16). Denoting the critical polar angle at the throat by θ_{th}, we exhibit θ_{th} also in Figure 4 versus the scaled coupling constant α/M^2 for the same sets of solutions for the GB invariant (c) and the CS invariant (d). The critical angle reaches its maximal value of about $\theta_{\mathrm{th,max}} \approx 0.19$ (c) and ≈ 0.18 (d) on the boundary of singular wormhole solutions. Along the boundary of degenerate wormholes the critical angle decreases monotonically, until it vanishes, precisely when the Schwarzschild-NUT solution is reached.

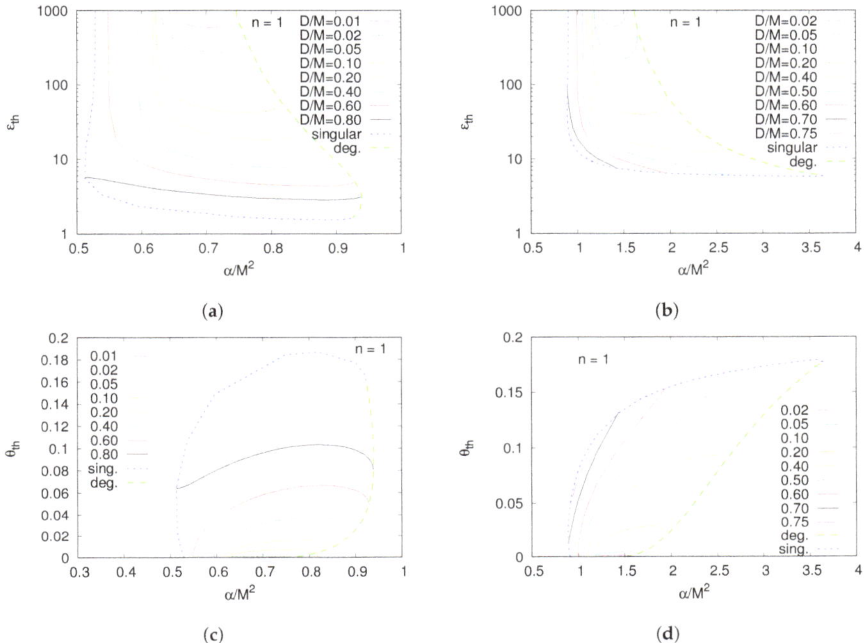

Figure 4. Properties at the throat (left plots: Gauss–Bonnet, right plots: Chern–Simons) for scaled NUT charge $n = N/M = 1$ and several values of the scaled scalar charge D/M: (**a**,**b**) energy density ϵ_{th}, (**c**,**d**) critical polar angle θ_{th} vs scaled coupling constant α/M^2. The dashed green curves represent the degenerate wormhole limit, and the dotted blue curves the singular limit.

The horizon metric does not feature a finite critical polar angle for the Schwarzschild-NUT and the scalarized nutty black holes. For these black holes $\theta_{\text{th}} = 0$ only on the polar axis, $\theta = 0$ and $\theta = \pi/2$. Consequently, $\theta_{\text{th}} = 0$ along the black hole boundary. Along the degenerate wormhole boundary θ_{th} then increases again. Thus all scalarized nutty wormholes do possess a finite critical polar angle. The region of closed timelike curves therefore extends to the throat of these nutty wormholes.

4. Conclusions

Following the reasoning of Brihaye et al. [46], who have studied spontaneously scalarized Schwarzschild-NUT solutions, whose scalarization is caused by the presence of either a GB term or a CS term in the scalar field equation, we have investigated scalarized nutty wormhole solutions in these higher curvature theories. The presence of a NUT charge leads to solutions with a Misner string on the polar axis. However, the dependence of the polar angle factorizes and thus only a set of coupled ODEs for the metric functions and the scalar field arises. Moreover the usual boundary conditions at spatial infinity are retained.

Solving numerically the ODEs we have obtained scalarized nutty wormhole solutions for both higher curvature invariants. These wormhole solutions possess a minimum of the circumferential radius, that arises naturally in these solutions, and that is identified with the wormhole throat. When integrating the ODEs beyond the throat, a maximum may be encountered, that would correspond to an equator. In any case, when integrating further into the second part of the manifold a singularity would be encountered at some value of the radial coordinate. To avoid such a singularity, we have imposed reflection symmetry on the wormhole solutions at the throat, and satisfied the resulting junction conditions in terms of an action with a thin shell of ordinary matter at the throat.

The novel scalarized GB wormholes represent a NUT generalization of the previously obtained scalarized wormholes [59–61,65]. In contrast, scalarized CS wormholes were not studied before, since a finite CS term is necessary for their existence, while a spherically symmetric metric leads to a vanishing CS invariant. Nevertheless the scalarized nutty wormholes with both invariants show many features already known from the ordinary scalarized GB wormholes. In particular, their domains of existence contain the same type of boundaries. These boundaries consist of the respective black hole boundary, of a boundary, where singular solutions are reached, and of a degenerate boundary, where two extrema, the equator and the throat, merge. However, the NUT charge gives rise to regions of the spacetime, where closed timelike curves exist, that extend to the throat.

Considering the limit of vanishing coupling constant, the Schwarzschild-NUT solutions are obtained for both invariants. Then wormhole solutions do not exist any longer, since they need the higher order curvature terms in the (generalized) Einstein equations to obtain an effective stress energy tensor that violates the NEC conditions. Considering on the other hand the limit of vanishing NUT charge, for both invariants a finite domain of existence of scalarized solutions results. For the GB term this is no surprise, since their existence was shown before [59–61,65]. For the CS, however, this limit is special. Scalarization needs higher and higher values of the coupling constant, as the NUT charge is lowered towards zero. In the limit, the coupling constant diverges as the NUT charge goes to zero. In that case, the CS term approaches a finite limiting value, which gives rise to a finite limiting domain of existence. Note that the spacetime of these wormhole and black hole solutions do not possess a Misner string anymore and consequently no closed timelike curves exist.

The presence of a NUT charge can be viewed as toy model for learning about scalarized rotating wormhole solutions. Their construction still represents a challenging task, not only because complicated sets of coupled partial differential equations need to be solved, but also because the proper conditions for the throat need to be formulated together with the proper set of junction conditions. However, further directions of research look promising like, in particular, the inclusion of further fields (see e.g., [70,71]), or the investigation of EGB theories viable in cosmological settings also for non-vanishing scalar field (see e.g., [72,73].)

Author Contributions: All the authors contributed equally to the conceptualization, methodology, software, validation, formal analysis, investigation, resources, data curation, original draft preparation, review and editing, visualization, supervision, project administration, funding acquisition. All authors have read and agreed to the published version of the manuscript.

Funding: This research was funded by: DFG Research Training Group 1620 *Models of Gravity*; COST Action CA16104.

Institutional Review Board Statement: Not applicable.

Informed Consent Statement: Not applicable.

Data Availability Statement: The data presented in this study are available on request from the corresponding author.

Acknowledgments: B.K. and J.K. gratefully acknowledge support by the DFG Research Training Group 1620 *Models of Gravity* and the COST Action CA16104.

Conflicts of Interest: The authors declare no conflicts of interest.

Appendix A

With $R_c = r e^{f_1/2}$, the definition of the throat $dR_c/dr(r_0) = 0$, $d^2R_c/dr^2(r_0) < 0$, where r_0 denotes the location of the throat, yields

$$\frac{1}{2}e^{f_1/2}\left(f_1' + 2\right) = 0\,,$$

$$\frac{1}{2r_0}e^{f_1/2}\left(f_1'' - 2\right) > 0\,,$$

where all quantities are evaluated at $r = r_0$ and the prime denotes the derivative with respect to the dimensionless variable $x = r/r_0$. f_1'' can be expressed in terms of lower order derivatives of the metric functions and the scalar field. This yields for the coupling to the Gauss–Bonnet term

$$
\begin{aligned}
R_C'' \;=\; & 128\alpha^2 e^{3f_1/2}\phi^2 r_0 \left[-6N^4 e^{2f_0}\left(3(f_0')^2 - 2(\phi_0')^2\right) - N^2 e^{f_0}e^{f_1}r_0^2\left(6(f_0')^2 - 7(\phi_0')^2\right) + e^{2f_1}(\phi_0')^2 r_0^4\right] \\
& + 16\alpha e^{5f_1/2}\phi_0' r_0^5\left[3e^{f_0}f_0'\phi N^2 - 6e^{f_0}\phi_0' N^2 + e^{f_1}f_0'\phi r_0^2 - 2e^{f_1}\phi_0' r_0^2\right] \\
& + e^{9f_1/2}r_0^7\left[12e^{f_0}N^2 - e^{f_1}(\phi_0')^2 r_0^2 + 4e^{f_1}r_0^2\right] \\
& \times \left\{2048\alpha^3 f_0'\phi^3\phi_0'\left[9e^{2f_0}N^4 + 6e^{f_0}e^{f_1}r_0^2 N^2 + e^{2f_1}r_0^4\right]\right. \\
& \left. - 512\alpha^2 e^{f_1}\phi^2 r_0^2\left[21e^{2f_0}N^4 + 10e^{f_0}e^{f_1}r_0^2 N^2 + e^{2f_1}r_0^4\right] + 8e^{5f_1}r_0^{10}\right\}^{-1} \\
> {}& 0 ,
\end{aligned}
\tag{A1}
$$

where f_0' and ϕ' are related by the constraint

$$
0 \;=\; 16\alpha f_0'\phi\phi_0'\left[3e^{f_0}N^2 + e^{f_1}r_0^2\right] + e^{f_1}r_0^2 - \left[4e^{f_0}N^2 + e^{f_1}(\phi_0')^2 r_0^2 + 4e^{f_1}r_0^2\right] .
\tag{A2}
$$

Similarly, for the coupling to the Chern–Simons term we find

$$
\begin{aligned}
R_C'' \;=\; & \Big(1024\alpha^3 e^{3f_0/2}(f_0')^4\phi^3\phi' N^3 \\
& - \alpha^2 e^{f_0}e^{3f_1/2}N^2\left[1280e^{f_0}e^{-f_1}(f_0')^2\phi^2 r_0 N^2 + 192(f_0')^2\phi^2(\phi')^2 r_0^3 - 256(f_0')^2\phi^2 r_0^3\right] \\
& + \alpha e^{f_0/2}e^{3f_1}N\left[16(f_0')^2\phi\phi' r_0^6 - 32f_0'(\phi')^2 r_0^6 - 352e^{f_0}e^{-f_1}\phi\phi' r_0^4 N^2 + 8\phi(\phi')^3 r_0^6 - 160\phi\phi' r_0^6\right] \\
& - 12e^{f_0}e^{7f_1/2}r_0^7 N^2 + e^{9f_1/2}(\phi')^2 r_0^9 - 4e^{9f_1/2}r_0^9\Big) \\
& \times \left\{8\left[192\alpha^2 e^{f_0}(f_0')^2\phi^2 N^2 - 24\alpha e^{f_0/2}e^{3f_1/2}\phi(\phi')r_0^3 N - e^{3f_1}r_0^6\right]e^{f_1}r_0^4\right\}^{-1} \\
> {}& 0 ,
\end{aligned}
\tag{A3}
$$

and the constraint

$$
0 = \alpha e^{f_0/2}(f_0')^2\phi\phi' N + 4e^{f_0}e^{f_1/2}r_0 N^2 + e^{3f_1/2}(\phi')^2 r_0^3 + 4e^{3f_1/2}r_0^3 .
\tag{A4}
$$

We note from Figure 3f that $\phi \geq 0$ at the throat. As a consequence of the constraint we find $\phi' \leq 0$ at the throat.

References

1. Damour, T.; Esposito-Farese, G. Nonperturbative strong field effects in tensor—Scalar theories of gravitation. *Phys. Rev. Lett.* **1993**, *70*, 2220.
2. Antoniou, G.; Bakopoulos, A.; Kanti, P. Evasion of No-Hair Theorems and Novel Black-Hole Solutions in Gauss–Bonnet Theories. *Phys. Rev. Lett.* **2018**, *120*, 131102.
3. Doneva, D.D.; Yazadjiev, S.S. New Gauss–Bonnet black holes with curvature induced scalarization in the extended scalar-tensor theories. *Phys. Rev. Lett.* **2018**, *120*, 131103.
4. Silva, H.O.; Sakstein, J.; Gualtieri, L.; Sotiriou, T.P.; Berti, E. Spontaneous scalarization of black holes and compact stars from a Gauss–Bonnet coupling. *Phys. Rev. Lett.* **2018**, *120*, 131104.
5. Antoniou, G.; Bakopoulos, A.; Kanti, P. Black-Hole Solutions with Scalar Hair in Einstein-Scalar–Gauss–Bonnet Theories. *Phys. Rev. D* **2018**, *97*, 084037.
6. Blázquez-Salcedo, J.L.; Doneva, D.D.; Kunz, J.; Yazadjiev, S.S. Radial perturbations of the scalarized Einstein-Gauss–Bonnet black holes. *Phys. Rev. D* **2018**, *98*, 084011.

7. Doneva, D.D.; Kiorpelidi, S.; Nedkova, P.G.; Papantonopoulos, E.; Yazadjiev, S.S. Charged Gauss–Bonnet black holes with curvature induced scalarization in the extended scalar-tensor theories. *Phys. Rev. D* **2018**, *98*, 104056.

8. Minamitsuji, M.; Ikeda, T. Scalarized black holes in the presence of the coupling to Gauss–Bonnet gravity. *Phys. Rev. D* **2019**, *99*, 044017.

9. Silva, H.O.; Macedo, C.F.B.; Sotiriou, T.P.; Gualtieri, L.; Sakstein, J.; Berti, E. Stability of scalarized black hole solutions in scalar–Gauss–Bonnet gravity. *Phys. Rev. D* **2019**, *99*, 064011.

10. Brihaye, Y.; Ducobu, L. Hairy black holes, boson stars and non-minimal coupling to curvature invariants. *Phys. Lett. B* **2019**, *795*, 135.

11. Myung, Y.S.; Zou, D. Quasinormal modes of scalarized black holes in the Einstein–Maxwell–Scalar theory. *Phys. Lett. B* **2019**, *790*, 400–407.

12. Bakopoulos, A.; Antoniou, G.; Kanti, P. Novel Black-Hole Solutions in Einstein-Scalar–Gauss–Bonnet Theories with a Cosmological Constant. *Phys. Rev. D* **2019**, *99*, 064003.

13. Doneva, D.D.; Staykov, K.V.; Yazadjiev, S.S. Gauss–Bonnet black holes with a massive scalar field. *Phys. Rev. D* **2019**, *99*, 104045.

14. Myung, Y.S.; Zou, D.C. Black holes in Gauss–Bonnet and Chern–Simons-scalar theory. *Int. J. Mod. Phys. D* **2019**, *28*, 1950114.

15. Macedo, C.F.B.; Sakstein, J.; Berti, E.; Gualtieri, L.; Silva, H.O.; Sotiriou, T.P. Self-interactions and Spontaneous Black Hole Scalarization. *Phys. Rev. D* **2019**, *99*, 104041.

16. Cunha, P.V.P.; Herdeiro, C.A.R.; Radu, E. Spontaneously Scalarized Kerr Black Holes in Extended Scalar-Tensor-Gauss–Bonnet Gravity. *Phys. Rev. Lett.* **2019**, *123*, 011101.

17. Bakopoulos, A.; Kanti, P.; Pappas, N. Existence of solutions with a horizon in pure scalar–Gauss–Bonnet theories. *Phys. Rev. D* **2020**, *101*, 044026.

18. Hod, S. Spontaneous scalarization of Gauss–Bonnet black holes: Analytic treatment in the linearized regime. *Phys. Rev. D* **2019**, *100*, 064039.

19. Collodel, L.G.; Kleihaus, B.; Kunz, J.; Berti, E. Spinning and excited black holes in Einstein-scalar–Gauss–Bonnet theory. *Class. Quant. Grav.* **2020**, *37*, 075018.

20. Bakopoulos, A.; Kanti, P.; Pappas, N. Large and Ultra-compact Gauss–Bonnet Black Holes with a Self-interacting Scalar Field. *Phys. Rev. D* **2020**, *101*, 084059.

21. Blázquez-Salcedo, J.L.; Doneva, D.D.; Kahlen, S.; Kunz, J.; Nedkova, P.; Yazadjiev, S.S. Axial perturbations of the scalarized Einstein-Gauss–Bonnet black holes. *Phys. Rev. D* **2020**, *101*, 104006.

22. Blázquez-Salcedo, J.L.; Doneva, D.D.; Kahlen, S.; Kunz, J.; Nedkova, P.; Yazadjiev, S.S. Polar quasinormal modes of the scalarized Einstein-Gauss–Bonnet black holes. *arXiv* **2020**, arXiv:2006.06006.

23. Herdeiro, C.A.R.; Radu, E.; Silva, H.O.; Sotiriou, T.P.; Yunes, N. Spin-induced scalarized black holes. *arXiv* **2020**, arXiv:2009.03904.

24. Berti, E.; Collodel, L.G.; Kleihaus, B.; Kunz, J. Spin-induced black-hole scalarization in Einstein-scalar–Gauss–Bonnet theory. *arXiv* **2020**, arXiv:2009.03905.

25. Dima, A.; Barausse, E.; Franchini, N.; Sotiriou, T.P. Spin-induced black hole spontaneous scalarization. *Phys. Rev. Lett.* **2020**, *125*, 231101.

26. Hod, S. Onset of spontaneous scalarization in spinning Gauss–Bonnet black holes. *Phys. Rev. D* **2020**, *102*, 084060.

27. Doneva, D.D.; Collodel, L.G.; Krüger, C.J.; Yazadjiev, S.S. Black hole scalarization induced by the spin: 2+1 time evolution. *Phys. Rev. D* **2020**, *102*, 104027.

28. Ayzenberg, D.; Yagi, K.; Yunes, N. Linear Stability Analysis of Dynamical Quadratic Gravity. *Phys. Rev. D* **2014**, *89*, 044023.

29. Kobayashi, T.; Motohashi, H.; Suyama, T. Erratum: Black hole perturbation in the most general scalar-tensor theory with second-order field equations I: The odd-parity sector. *Phys. Rev. D* **2012**, *85*, 084025. *Phys. Rev. D* **2017**, *96*, 109903.

30. Kobayashi, T.; Motohashi, H.; Suyama, T. Black hole perturbation in the most general scalar-tensor theory with second-order field equations II: The even-parity sector. *Phys. Rev. D* **2014**, *89*, 084042.

31. Yunes, N.; Pretorius, F. Dynamical Chern–Simons Modified Gravity. I. Spinning Black Holes in the Slow-Rotation Approximation. *Phys. Rev. D* **2009**, *79*, 084043.

32. Konno, K.; Matsuyama, T.; Tanda, S. Rotating black hole in extended Chern–Simons modified gravity. *Prog. Theor. Phys.* **2009**, *122*, 561.
33. Cambiaso, M.; Urrutia, L.F. An extended solution space for Chern–Simons gravity: The slowly rotating Kerr black hole. *Phys. Rev. D* **2010**, *82*, 101502.
34. Yagi, K.; Yunes, N.; Tanaka, T. Erratum: Slowly Rotating Black Holes in Dynamical Chern–Simons Gravity: Deformation Quadratic in the Spin. *Phys. Rev. D* **2012**, *86*, 044037. *Phys. Rev. D* **2014**, *89*, 049902.
35. Stein, L.C. Rapidly rotating black holes in dynamical Chern–Simons gravity: Decoupling limit solutions and breakdown. *Phys. Rev. D* **2014**, *90*, 044061.
36. Konno, K.; Takahashi, R. Scalar field excited around a rapidly rotating black hole in Chern–Simons modified gravity. *Phys. Rev. D* **2014**, *90*, 064011.
37. McNees, R.; Stein, L.C.; Yunes, N. Extremal black holes in dynamical Chern–Simons gravity. *Class. Quant. Grav.* **2016**, 33, 235013.
38. Delsate, T.; Herdeiro, C.; Radu, E. Non-perturbative spinning black holes in dynamical Chern–Simons gravity. *Phys. Lett. B* **2018**, *787*, 8.
39. Taub, A.H. Empty space-times admitting a three parameter group of motions. *Annals Math.* **1951**, *53*, 472.
40. Newman, E.T.; Tamburino, L.; Unti, T. Empty space generalization of the Schwarzschild metric. *J. Math. Phys.* **1963**, *4*, 915.
41. Misner, C.W. Taub-NUT space as a counterexample to almost anything. *Relat. Theory Astrophys.* **1967**, *1*, 160.
42. Misner, C.W. *Relativity Theory and Astrophysics I: Relativity and Cosmology*; Ehlers, J., Ed.; Lectures in Applied Mathematics; American Mathematical Society: Providence, RI, USA, 1967; Volume 8, p. 160.
43. Bonnor, W.B. A new interpretation of the NUT metric in general relativity. *Proc. Camb. Phil. Soc.* **1969**, *66*, 145.
44. Bonnor, W.B. The interactions between two classical spinning particles. *Class. Quant. Grav.* **2001**, *18*, 1381.
45. Manko, V.S.; Ruiz, E. Physical interpretation of NUT solution. *Class. Quant. Grav.* **2005**, *22*, 3555.
46. Brihaye, Y.; Herdeiro, C.; Radu, E. The scalarised Schwarzschild-NUT spacetime. *Phys. Lett. B* **2019**, *788*, 295.
47. Brihaye, Y.; Radu, E. Remarks on the Taub-NUT solution in Chern–Simons modified gravity. *Phys. Lett. B* **2017**, *764*, 300.
48. Morris, M.S.; Thorne, K.S. Wormholes in space-time and their use for interstellar travel: A tool for teaching general relativity. *Am. J. Phys.* **1988**, *56*, 395.
49. Visser, M. *Lorentzian Wormholes: From Einstein to Hawking*; AIP: Woodbury, NY, USA, 1995.
50. Lobo, F.S. Introduction. *Fundam. Theor. Phys.* **2017**, *189*, 1.
51. Visser, M.; Kar, S.; Dadhich, N. Traversable wormholes with arbitrarily small energy condition violations. *Phys. Rev. Lett.* **2003**, *90*, 201102.
52. Kar, S.; Dadhich, N.; Visser, M. Quantifying energy condition violations in traversable wormholes. *Pramana* **2004**, *63*, 859.
53. Hochberg, D. Lorentzian wormholes in higher order gravity theories. *Phys. Lett.* **1990**, *B251*, 349.
54. Fukutaka, H.; Tanaka, K.; Ghoroku, K. Wormhole Solutions In Higher Derivative Gravity. *Phys. Lett. B* **1989**, *222*, 191.
55. Ghoroku, K.; Soma, T. Lorentzian wormholes in higher derivative gravity and the weak energy condition. *Phys. Rev.* **1992**, *D46*, 1507.
56. Furey, N.; De Benedictis, A. Wormhole throats in R^m gravity. *Class. Quant. Grav.* **2005**, *22*, 313.
57. Eiroa, E.F.; Richarte, M.G.; Simeone, C. Thin-shell wormholes in Brans-Dicke gravity. *Phys. Lett. A* **2008**, *373*, 1.
58. Bronnikov, K.A.; Elizalde, E. Spherical systems in models of nonlocally corrected gravity. *Phys. Rev. D* **2010**, *81*, 044032.
59. Antoniou, G.; Bakopoulos, A.; Kanti, P.; Kleihaus, B.; Kunz, J. Novel Einstein-scalar–Gauss–Bonnet wormholes without exotic matter. *Phys. Rev. D* **2020**, *101*, 024033.
60. Kanti, P.; Kleihaus, B.; Kunz, J. Wormholes in Dilatonic Einstein-Gauss–Bonnet Theory. *Phys. Rev. Lett.* **2011**, *107*, 271101.
61. Kanti, P.; Kleihaus, B.; Kunz, J. Stable Lorentzian Wormholes in Dilatonic Einstein-Gauss–Bonnet Theory. *Phys. Rev. D* **2012**, *85*, 044007.
62. Lobo, F.S.N.; Oliveira, M.A. Wormhole geometries in f(R) modified theories of gravity. *Phys. Rev. D* **2009**, 104012.

63. Harko, T.; Lobo, F.S.N.; Mak, M.K.; Sushkov, S.V. Modified-gravity wormholes without exotic matter. *Phys. Rev. D* **2013**, *87*, 067504.

64. Kanti, P.; Mavromatos, N.E.; Rizos, J.; Tamvakis, K.; Winstanley, E. Dilatonic Black Holes in Higher Curvature String Gravity. *Phys. Rev. D* **1996**, *54*, 5049.

65. Ibadov, R.; Kleihaus, B.; Kunz, J.; Murodov, S. Wormholes in Einstein-scalar–Gauss–Bonnet theories with a scalar self-interaction potential. *Phys. Rev. D* **2020**, *102*, 064010.

66. Israel, W. Erratum: Singular hypersurfaces and thin shells in general relativity. *Nuovo Cim. B* **1966**, *44S10*, 1. *Nuovo Cim. B* **1967**, *48*, 463.

67. Davis, S.C. Generalized Israel junction conditions for a Gauss–Bonnet brane world. *Phys. Rev. D* **2003**, *67*, 024030.

68. Kleihaus, B.; Kunz, J.; Kanti, P. Particle-like ultracompact objects in Einstein-scalar–Gauss–Bonnet theories. *Phys. Lett. B* **2020**, *804*, 135401.

69. Kleihaus, B.; Kunz, J.; Kanti, P. Properties of ultracompact particlelike solutions in Einstein-scalar–Gauss–Bonnet theories. *Phys. Rev. D* **2020**, *102*, 024070.

70. Clément, G.; Gal'tsov, D.; Guenouche, M. NUT wormholes. *Phys. Rev. D* **2016**, *93*, 024048.

71. Brihaye, Y.; Renaux, J. Scalarized-charged wormholes in Einstein-Gauss–Bonnet gravity. *arXiv* **2020**, arXiv:2004.12138.

72. Odintsov, S.D.; Oikonomou, V.K. Inflationary Phenomenology of Einstein Gauss–Bonnet Gravity Compatible with GW170817. *Phys. Lett. B* **2019**, *797*, 134874.

73. Oikonomou, V.K.; Fronimos, F.P. Reviving non-Minimal Horndeski-like Theories after GW170817: Kinetic Coupling Corrected Einstein-Gauss–Bonnet Inflation. *arXiv* **2020**, arXiv:2006.05512.

Article

Relativistic Symmetries and Hamiltonian Formalism

Piotr Kosiński * and Paweł Maślanka *

Faculty of Physics and Applied Informatics, University of Lodz, Narutowicza 68, 90-136 Lodz, Poland
* Correspondence: piotr.kosinski@uni.lodz.pl (P.K.); pawel.maslanka@uni.lodz.pl (P.M.)

Received: 30 September 2020; Accepted: 29 October 2020; Published: 1 November 2020

Abstract: The relativistic (Poincaré and conformal) symmetries of classical elementary systems are briefly discussed and reviewed. The main framework is provided by the Hamiltonian formalism for dynamical systems exhibiting symmetry described by a given Lie group. The construction of phase space and canonical variables is given using the tools from the coadjoint orbits method. It is indicated how the "exotic" Lorentz transformation properties for particle coordinates can be derived; they are shown to be the natural consequence of the formalism.

Keywords: coadjoint orbits; conformal group; Poincaré group

1. Introduction

We present here a brief review of some results obtained by our colleagues in collaboration with Yves Brihaye and us. They concern the old topic of the basic role of space-time symmetries in physics. It appears that some of the main results, which were for the first time obtained and developed in quantum theory, can be described quite precisely on the classical level in the framework of Hamiltonian formalism. Then, their quantum counterparts are recovered by applying a straightforward canonical quantization procedure. The problem is not only academic. For example, in recent years, much effort has been devoted to exploring the anomaly-related phenomena in kinetic theory [1,2]. This work is to a great extent based on semiclassical approximation and the description of various symmetries within this approximation. The sound knowledge concerning the (semi)classical aspects of relativistic symmetries contributes to a deeper understanding of such phenomena.

Part of the results reported here has been obtained in collaboration with our longtime friend Yves Brihaye. It was always a great pleasure to work with him.

2. Orbit Method

In physics, beauty is often identified with symmetry. Having at our disposal two theories explaining the same set of experimental data, we are inclined to choose the one exhibiting more symmetry. This strategy is somehow supported by our experience. A dazzling example is the success of general relativity, which is confirmed over and over again by experiments, observations and even everyday life. Although some physicists warn that beauty can lead one astray [3], we are, generally speaking, attached to the idea that adopting symmetry as a guiding principle often leads to the theories with considerable predictive power.

The symmetry of the physical system is described, at the formal level, by the choice of symmetry group G. In quantum theory, the states of the physical system are classified according to the unitary representations of G. However, even if the group G represents the maximal symmetry of some system, there is still much freedom in the choice of the total space of states and relevant observables. The exception is provided by the so-called elementary systems, which are, by definition, described by irreducible representations of the symmetry group. For the elementary system, given the symmetry group G, one can classify all admissible spaces of states and construct all relevant observables in purely group-theoretical terms. For example, an elementary relativistic particle is described by an unitary irreducible representation of the Poincaré group [4,5]. All its states are explicitly known and may be obtained by acting with group elements on some fixed state; all observables, like energy, momentum and angular momentum (and coordinate), are constructed out of group generators. The same concerns nonrelativistic particle; the relevant group here is the (quantum mechanical) Galilei group [6].

Let us note that, in order to construct the quantum description of elementary particles, we do not have to refer to classical theory, canonical quantization, etc. However, it would be nice to find if the notion of elementary system, formulated in group-theoretical language, can be extended to classical physics in such a way that the canonical quantization of the latter yields the corresponding quantum elementary system. This question has been considered by a number of authors. The mathematical basis has been laid out by Kirillov [7,8], who developed the so-called orbit method. Souriau [9] elaborated symplectic aspects of classical and quantum physics.

Let us sketch the main points of the description of the classical system exhibiting symmetry [10,11]. The general framework for classical dynamics is provided by the Hamiltonian formalism. The space of states is a symplectic manifold—a phase space. The symplectic structure allows one to define the Poisson bracket. Once a Hamiltonian—some function on the phase space—is selected, one can write the canonical equations of motion, which determine the actual dynamics. The transformations preserving symplectic structures are called the canonical transformations. The symmetry transformations are the canonical transformations which preserve the functional form of the Hamiltonian. The set of symmetry transformations form a group, which is the symmetry group of a given dynamical system. The classical system is called elementary if the symmetry group acts transitively on phase space. It appears that in this case, the Hamiltonian formalism can be described in group-theoretical terms. To this end, let G be a Lie group with Lie algebra

$$[X_\alpha, X_\beta] = i c_{\alpha\beta}^{\ \ \gamma} X_\gamma. \tag{1}$$

In the Lie algebra, G acts through adjoint representation

$$Ad_g(X_\alpha) = g X_\alpha g^{-1} = D^\beta_{\ \alpha}(g) X_\beta. \tag{2}$$

In the dual space to the Lie algebra, there acts the contragradient representation called the coadjoint one. Let ζ_α be the coordinates in the dual space; then

$$Ad_g^*(\zeta_\alpha) = D^\beta_{\ \alpha}(g^{-1}) \zeta_\beta. \tag{3}$$

The orbit of coadjoint action (3) is called the coadjoint orbit.

In the dual space, one can define a natural Poisson structure

$$\{\zeta_\alpha, \zeta_\beta\} = c_{\alpha\beta}^{\ \ \gamma} \zeta_\gamma. \tag{4}$$

This Poisson structure is, in general, degenerate, i.e., there exist nonconstant functions having vanishing Poisson brackets with all ζ_α. However, the important point is that the Poisson brackets (4) can be consistently restricted to the orbits, and then they become nondegenerate; the orbits are symplectic manifolds. Moreover, if G acts transitively on phase space, then, modulo some mild (i.e., fulfilled in most physical contexts) assumptions, the phase space can be identified with some coadjoint orbit of G. Therefore, for the elementary systems, the phase space together with its symplectic structure are described in purely group-theoretical terms. All observables, being the functions over the phase space, are expressible in terms of the coordinates ζ_α. One can view them as classical counterparts of the elements of enveloping algebra of the Lie algebra under consideration. What remains is the choice of the Hamiltonian, which defines the actual physical system. If G describes the space-time symmetry, one of its generators is a natural candidate for the Hamiltonian. Then, the group-theoretical description is complete: all elements of Hamiltonian formalism are expressed in terms of group notions.

The first step to classify the elementary systems with a given symmetry group is to characterize the set of coadjoint orbits. The generic orbits are the submanifolds of the dual space to Lie algebra obtained by fixing the values of all functionally independent Casimir functions (which are classical counterparts of the Casimir operators). However, there exist also nongeneric orbits of lower dimension (the extreme case being the trivial orbit). They are characterized by an additional relation, which can be found as follows. The infinitesimal action of the symmetry group on coadjoint orbit/phase space is given by the Poisson bracket with a relevant generator. Therefore, if we nullify some ideal in the Poisson algebra of functions over phase space, the resulting relations will be invariant under the coadjoint action of G. This is quite a convenient way of characterizing nongeneric orbits. Once the orbit is explicitly characterized, the next step is to find convenient coordinates (Darboux variables). Finally, we have to write out the Hamiltonian in terms of independent canonical variables, which completes the description.

3. Poincaré Symmetry

In an attempt to understand the origin of quantum mechanics of spinning particle (in particular, spin), various classical models have been proposed [12–37]. Let us sketch the description based on the ideas presented in the previous section [38].

The relativistic symmetry is described by the Poincaré group. It consists of Lorentz transformations Λ ($\Lambda^T \eta \Lambda = \eta$, $\eta = \mathrm{diag}(+ - - -)$) and translations a. The composition law reads

$$(\Lambda, a) \cdot (\Lambda', a') = (\Lambda\Lambda', \Lambda a' + a). \tag{5}$$

The relevant Lie algebra consists of Lorentz ($M_{\mu\nu}$) and translation (P_μ) generators obeying

$$[M_{\mu\nu}, P_\alpha] = i(\eta_{\nu\alpha} P_\mu - \eta_{\mu\alpha} P_\nu), \tag{6}$$

$$[M_{\mu\nu}, M_{\alpha\beta}] = i(\eta_{\mu\beta} M_{\nu\alpha} + \eta_{\nu\alpha} M_{\mu\beta} - \eta_{\mu\alpha} M_{\nu\beta} - \eta_{\nu\beta} M_{\mu\alpha}), \tag{7}$$

$$[P_\mu, P_\nu] = 0. \tag{8}$$

There are two Casimir operators, mass and spin,

$$M^2 \equiv P^\mu P_\mu, \tag{9}$$

$$W^2 \equiv W^\mu W_\mu, \quad W^\mu = \frac{1}{2}\epsilon^{\mu\nu\alpha\beta} P_\nu M_{\alpha\beta}. \tag{10}$$

The relevant coordinates in the dual space to Lie algebra are ζ_μ and $\zeta_{\mu\nu} = -\zeta_{\nu\mu}$. For physical reasons, we are interested in the case $M^2 \geqslant 0$; then, the coadjoint orbits consist of two disjoint pieces corresponding to $\zeta_0 \gtrless 0$; since ζ_0, being the counterpart of time translation generator P_0, represents energy, we restrict ourselves to $\zeta_0 > 0$ case.

Now, the generic orbits are obtained by fixing M^2 and W^2. Therefore, they are eight-dimensional. This is exactly what we expect: there are three components of position and momentum together with two variables describing the spin of fixed length (due to fixing W^2). Explicitly,

$$m^2 = \zeta_\mu \zeta^\mu, \tag{11}$$

$$-m^2 s^2 = w^\mu w_\mu, \quad w^\mu = \frac{1}{2}\epsilon^{\mu\nu\alpha\beta} \zeta_\nu \zeta_{\alpha\beta}. \tag{12}$$

Note that for $m^2 = 0$, the invariants are no longer independent. Therefore, we start with $m^2 > 0$. It can be shown that the canonical point on coadjoint orbit can be chosen as [38]:

$$\underline{\zeta}_\mu = (m, \vec{0}) \,, \tag{13}$$

$$\underline{\zeta}_{0i} = -\underline{\zeta}_{i0} = 0 \,, \tag{14}$$

$$\underline{\zeta}_{ij} = s\epsilon_{3ij} \,. \tag{15}$$

Any other point of the orbit is obtained from the canonical one by a coadjoint action of the Poincaré group. By an appropriate choice of parametrization, we find that the general point of the orbit can be written as [37,38]

$$\zeta_\mu = p_\mu \,, \tag{16}$$

$$\zeta_{0i} = -p_0 x_i + \frac{\epsilon_{ilk} s_l p_k}{m + p_0} \,, \tag{17}$$

$$\zeta_{ij} = x_i p_j - x_j p_i + s_k \epsilon_{kij} \,, \tag{18}$$

while the Kirillov symplectic structure yields

$$\{x_i, p_j\} = \delta_{ij} \,, \tag{19}$$

$$\{s_i, s_j\} = \epsilon_{ijk} s_k \,, \tag{20}$$

the remaining Poisson bracket being vanishing. It is quite straightforward to quantize the resulting symplectic structure. As a result, one obtains the well-known form of generators of unitary irreducible representations of the Poincaré group describing massive particles [39].

Assume now that $m^2 = 0$. Then, both invariants vanish independently of the value s. Moreover, in quantum theory of massless particles, spin is no longer a dynamical variable. In fact, quantum particles are uniquely characterized by chirality, which can be viewed as the projection of spin on momentum. However, it is a fixed number, not a dynamical variable. Therefore, we expect the phase space to be six-dimensional. The corresponding orbits must be nongeneric. There is only one independent Casimir function, so, according to the prescription given above, we have to construct the relevant ideal in the Poisson algebra. Consider the following functions on phase space.

$$I_\mu(\underline{\zeta}) \equiv w_\mu - s\zeta_\mu \,. \tag{21}$$

Then

$$\{I_\mu, \zeta_\nu\} = 0 \,, \tag{22}$$

$$\{I_\mu, \zeta_{\alpha\beta}\} = g_{\mu\alpha} I_\beta - g_{\mu\beta} I_\alpha \,, \tag{23}$$

so, one can put consistently $I_\mu = 0$. I_μ are the generators of invariant ideal. The relevant coadjoint orbit is defined by the equations .

$$\zeta^\mu \zeta_\mu = 0, \tag{24}$$

$$I_\nu = 0. \tag{25}$$

Equation (25) is called "enslaving" condition [40,41] (cf. also [9]). Due to $\zeta^\nu I_\nu = -s\zeta^\nu \zeta_\nu = 0$, only three Equations (25) are independent. Therefore, we obtain six-dimensional orbit. Again, it is not difficult to identify some canonical point on the orbit.

$$\underline{\zeta}_\mu = (k, 0, 0, k) \equiv k_\mu, \tag{26}$$

$$\underline{\zeta}_{\mu\nu} = \begin{cases} -s & , & (\mu\nu) = (12) \\ s & , & (\mu\nu) = (21) \\ 0 & , & (\mu\nu) \neq (12), (21). \end{cases} \tag{27}$$

By applying the coadjoint action of the Poincaré group and choosing suitable parametrization, we find [42]

$$\zeta_\mu = p_\mu, \tag{28}$$

$$\zeta_{0i} = -p_0 x_i, \tag{29}$$

$$\zeta_{ij} = x_i p_j - x_j p_i + s\frac{\epsilon_{ijk} p_k}{p_0}. \tag{30}$$

From Equation (30), we conclude that s is the projection of total angular momentum on the momentum direction.

Equations (28)–(30), when compared with the general form of Poisson brackets (4) and Equations (6)–(8), yield [42]

$$\{x_i, x_j\} = s\frac{\epsilon_{ijk} p_k}{p_0^3}, \tag{31}$$

$$\{x_i, p_j\} = \delta_{ij}. \tag{32}$$

with the remaining brackets vanishing. Note that the coordinates do not commute any longer (cf. (31)). This is because we are considering the nongeneric orbit defined by the additional ("enslaving") condition. This situation persists on the quantum level. A nice argument can be given [43,44] that it is not possible to define standard, i.e., commuting, coordinates for massless irreducible representation; only for reducible representation describing the helicities $S = \pm\frac{1}{2}$ such a coordinate operator exists [44].

Due to the more complicated form of Poisson brackets, the canonical quantization procedure is nontrivial. The coordinates cannot be represented by derivatives with respect to momentum; instead, a covariant derivative in the field of monopole must be used.

The approach described above provides a systematic way of studying symmetries on the level of classical Hamiltonian formalism. Phase space and dynamical observables are constructed in terms of group-theoretical notions and put on firm ground. In particular, one can study the transformation properties of various observables under the action of the symmetry group. This concerns, for example, the coordinate variable. It appears that it has "exotic" transformation properties. Consider, as an example, a massless particle with helicity s. The conserved generator of boosts reads (cf. Equation (29))

$$\zeta_{0i}(t) = p_0 x_i - p_i t. \tag{33}$$

By virtue of Equation (31), one finds [40–42,45,46]

$$\delta\vec{x} = \{\vec{x}, \delta v_i \zeta_{0i}(t)\} = -\delta\vec{v}t + \dot{\vec{x}}(\delta\vec{v} \cdot \vec{x}) + s\frac{\delta\vec{v} \times \vec{p}}{p_0^2}. \tag{34}$$

The first term on the right-hand side corresponds to the standard Lorentz transformation, while the second appears due to the fact that in the Hamiltonian formalism, time is kept fixed, so one has to recompute everything back to the initial time. The last term, which is helicity dependent, represents the so-called "side jump". The latter leads to the kinematical effect playing a role in impurity scattering caused by spin-orbit interaction [47] and relativistic Hall effect of light [48–54].

4. The Conformal Group

In four-dimensional space-time, the conformal group provides a fifteen-dimensional extension of the Poincaré group by scaling and special conformal transformations. It describes the approximate symmetry at energies large as compared to all dimensionful parameters. It would be interesting to provide the Hamiltonian description of all elementary conformally invariant dynamical systems.

The dual space to conformal Lie algebra is parametrized by ζ_μ, $\zeta_{\mu\nu}$ (Poincaré algebra), η (dilatations) and η_μ (special conformal transformations). Apart from the Poisson brackets following from commutation rules (6)–(8), we have the additional ones

$$\{\eta, \zeta_\mu\} = \zeta_\mu, \tag{35}$$

$$\{\eta, \zeta_{\mu\nu}\} = 0, \tag{36}$$

$$\{\eta, \eta_\mu\} = -\eta_\mu, \tag{37}$$

$$\{\zeta_{\mu\nu}, \eta_\rho\} = g_{\nu\rho}\eta_\mu - g_{\mu\rho}\eta_\nu, \tag{38}$$

$$\{\eta_\mu, \zeta_\nu\} = 2(\zeta_{\mu\nu} - g_{\mu\nu}\eta), \tag{39}$$

$$\{\eta_\mu, \eta_\nu\} = 0. \tag{40}$$

In order to classify the conformally invariant Hamiltonian system, we have to find all coadjoint orbits. It is not completely straightforward. The convenient way to do this is to use twistor formalism [55]. There are coadjoint orbits of dimensions 12 (generic), 10, 8 and 6. The latter case is particularly interesting: the Poincaré symmetry of quantum massless particles can be extended to the conformal one [56,57].

Therefore, we expect that the same holds true classically. In order to characterize six-dimensional orbits, one has to find nine independent generators of the relevant ideal. We already have four generators: $\zeta_\mu \zeta^\mu$ and I_μ (cf. Equation (21)). The remaining five can be chosen as [58]

$$J \equiv \eta + \frac{\zeta_k \zeta_{0k}}{\zeta_0} \, , \tag{41}$$

$$J_0 \equiv \eta_0 + \frac{\zeta_{0k}\zeta_{0k}}{\zeta_0} + \frac{\lambda^2}{\zeta_0} \, , \tag{42}$$

$$J_i \equiv \eta_i + \frac{\zeta_i \zeta_{0k}\zeta_{0k}}{\zeta_0^2} - 2\frac{\zeta_{0i}\zeta_k \zeta_{0k}}{\zeta_0^2} - 2s\epsilon_{ikl}\frac{\zeta_{0k}\zeta_l}{\zeta_0^2} - s^2 \frac{\zeta_i}{\zeta_0^2} \, . \tag{43}$$

Equations (41)–(43) merely imply that the generators of dilatations and special conformal transformations are functions on the phase space of Poincaré covariant massless particles

$$D = +p_k x_k \, , \tag{44}$$

$$K_i = -p_i x_k x_k + 2x_i x_k p_k - 2s\epsilon_{ikl}\frac{x_k p_l}{p_0} + \frac{s^2 p_i}{p_0^2} \, . \tag{45}$$

We conclude that, on the classical Hamiltonian level, the Poincaré symmetry of massless particles of arbitrary helicity may be extended to the conformal one. One can show [58] that the canonical quantization of the resulting structure can be performed in a more or less straightforward way, leading to the unitary irreducible representations of the conformal group belonging to the Mack list [57].

The dynamical systems corresponding to the orbits of higher dimensions will be described elsewhere [59].

5. Conclusions

Canonical quantization is the textbook method of constructing the quantum dynamics. One starts with some classical dynamical system defined within the Hamiltonian formalism and applies the canonical quantization procedure consisting in replacing the Poisson brackets by commutators (divided by $i\hbar$). With a little bit of luck, a consistent mathematical structure is obtained, which defines the quantum counterpart of the classical system we have started with.

However, the quantum dynamics should be the primary concept with the classical one emerging in the appropriate limit. Therefore, the question arises as to how to construct the quantum theory without referring to classical notions. One (the only?) way to do this is to refer to the symmetry arguments. According to the basic principles of quantum mechanics, the symmetry transformations are represented by unitary operators acting in the Hilbert space of states. Given a symmetry group G, all allowed spaces of states of the physical system can be classified and explicitly described, provided that the (projective) unitary representations of G are known. Furthermore, by identifying the elementary systems with irreducible representations one concludes that, in this case, all observables can be, at least in principle, constructed from the elements of the Lie algebra of G.

When such a construction is performed, one may consider (at least on a formal level) the limit $\hbar \to 0$, yielding some classical dynamics, and pose the question as to whether the former can be recovered by applying the canonical quantization to the latter. The answer to this question is affirmative in the case of relativistic space-time symmetries. We have sketched above the main steps of the relevant construction. The starting point is the notion of the elementary Hamiltonian system: it is the one exhibiting the symmetry which acts transitively on the phase space. Then, the candidates for the admissible phase spaces are provided by the coadjoint orbits of the symmetry group. It appears that both generic and nongeneric orbits should be considered; for example, the massless particles are described by nongeneric orbits of the Poincaré group. Once the orbit is selected, the explicit description of the resulting classical Hamiltonian system is obtained (one should keep in mind that for space-time symmetries, the Hamiltonian belongs to the Lie algebra of symmetry group).

Thus far, the above program has been completed for the Poincaré symmetry. The next step is to extend it to the conformal group. We have already shown [58] that the nongeneric six-dimensional coadjoint orbits of the conformal group describe massless particles; more precisely, in this case, the Poincaré symmetry can be extended to the conformal one. Canonical quantization (some care concerning the ordering problem must be exercised) of the generators of conformal group yields its unitary representation acting in the space of states of massless particles with fixed helicity; this result agrees with the one obtained by Mack [57]. It remains to consider the nongeneric orbits which are eight- and ten-dimensional as well as generic, twelve-dimensional ones. They should provide the classical counterparts of the remaining representations classified by Mack.

Author Contributions: Both authors contributed equally to this work. All authors have read and agreed to the published version of the manuscript.

Funding: This research received no external funding.

Acknowledgments: We are grateful to Joasia Gonera, Krzysztof Andrzejewski and Cezary Gonera for numerous, pleasant, and enlightening discussions.

Conflicts of Interest: The authors declare no conflict of interest.

References

1. Loganayagam, R.; Surówka, P. Anomaly/Transport in an Ideal Weyl Gas. *J. High Energy Phys.* **2012**, *1204*, 97. [CrossRef]
2. Son, D.T.; Yamamoto, N. Berry Curvature, Triangle Anomalies, and the Chiral Magnetic Effect in Fermi Liquids. *Phys. Rev. Lett.* **2012**, *109*, 181602. [CrossRef] [PubMed]
3. Hossenfelder, S. *Lost in Math: How Beauty Leads Physics Astray*; Basic Books: New York, NY, USA, 2018.
4. Wigner, E. On Unitary Representations of the Inhomogeneous Lorentz Group. *Ann. Math.* **1939**, *40*, 149–204. [CrossRef]
5. Weinberg, S. *The Quantum Theory of Fields I*; Cambridge University Press: Cambridge, UK, 1995.
6. Lévy-Leblond, J.M. Nonrelativistic Particles and Wave Equations. *Commun. Math. Phys.* **1967**, *6*, 286–311. [CrossRef]
7. Kirillov, A.A. *Elements of the Theory of Representations*; Springer: Berlin/Heidelberg, Germany, 1976.
8. Kirillov, A.A. *Lectures on the Orbit Method. Graduate Studies in Mathematics*; American Mathematical Society: Providence, RI, USA, 2004; Volume 64.
9. Souriau, J.M. *Structure of Dynamical Systems: A Symplectic View of Physics*; Birkhauser: Basel, Switzerland, 1997.
10. Arnol'd, V.I. *Mathematical Methods of Classical Mechanics*; Springer: Berlin/Heidelberg, Germany, 1989.
11. Marsden, J.E.; Ratiu, T.S. *Introduction to Mechanics and Symmetry*; Springer: Berlin/Heidelberg, Germany, 1999.
12. Frenkel, J. Die Elektrodynamik des Rotierenden Elektrons. *Z. Phys.* **1926**, *37*, 243–262. [CrossRef]
13. Thomas, L.H. The Motion of the Spinning Electron. *Nature* **1926**, *117*, 514. [CrossRef]
14. Thomas, L.H. The Kinematics of an Electron with an Axis. *Lond. Edinb. Dublin Philos. Mag. J. Sci.* **1927**, *3*, 1–22. [CrossRef]
15. Kramers, H. On the Classical Theory of the Spinning Electron. *Physica* **1934**, *1*, 825–828. [CrossRef]
16. Mathisson, M. Neue Mechanik Materieller Systeme. *Acta Phys. Pol.* **1937**, *6*, 163–2900.
17. Papapetrou, A. Spinning Test-Particles in General Relativity. I. *Proc. R. Soc. Lond. Ser. A Math. Phys. Sci.* **1951**, *209*, 248–258.
18. Corinaldesi, E.; Papapetrou, A. Spinning Test-Particles in General Relativity. II. *Proc. R. Soc. Lond. Ser. A Math. Phys. Sci.* **1951**, *209*, 259–268.
19. Dixon, W. A Covariant Multipole Formalism for Extended Test Bodies in General Relativity. *Il Nuovo Cim. (1955–1965)* **1964**, *34*, 317–339. [CrossRef]

20. Dixon, W. Description of Extended Bodies by Multipole Moments in Special Relativity. *J. Math. Phys.* **1967**, *8*, 1591–1605. [CrossRef]

21. Dixon, W. Dynamics of Extended Bodies in General Relativity. I. Momentum and Angular Momentum. *Proc. R. Soc. Lond. Ser. A Math. Phys. Sci.* **1970**, *314*, 499–527.

22. Bhabha, H.J.; Corben, H.C. General Classical Theory of Spinning Particles in a Maxwell Field. *Proc. R. Soc. Lond. Ser. A Math. Phys. Sci.* **1941**, *178*, 273–314.

23. Corben, H. Spin in Classical and Quantum Theory. *Phys. Rev.* **1961**, *121*, 1833–1839. [CrossRef]

24. Corben, H. Spin Precession in Classical Relativistic Mechanics. *Il Nuovo Cim. (1955–1965)* **1961**, *20*, 529–541. [CrossRef]

25. Nyborg, P. On Classical Theories of Spinning Particles. *Il Nuovo Cim. (1955–1965)* **1962**, *23*, 47. [CrossRef]

26. Frydryszak, A. Lagrangian Models of Particles with Spin: The First Seventy Years. In *From Field Theory to Quantum Groups*; World Scientific: Singapore, 1996; pp. 151–172.

27. Gaioli, F.H.; Alvarez, E.T.G. Classical and Quantum Theories of Spin. *Found. Phys.* **1998**, *28*, 1539–1550. [CrossRef]

28. Deriglazov, A.A. Variational Problem for the Frenkel and the Bargmann–Michel–Telegdi (BMT) Equations. *Mod. Phys. Lett. A* **2013**, *28*, 1250234. [CrossRef]

29. Deriglazov, A.A. Lagrangian for the Frenkel Electron. *Phys. Lett. B* **2014**, *736*, 278–282. [CrossRef]

30. Costa, L.F.; Herdeiro, C.; Natário, J.; Zilhao, M. Mathisson's Helical Motions for a Spinning Particle: Are They Unphysical? *Phys. Rev. D* **2012**, *85*, 024001. [CrossRef]

31. Costa, L.F.; Natário, J. Center of Mass, Spin Supplementary Conditions, and the Momentum of Spinning Particles. In *Equations of Motion in Relativistic Gravity*; Springer: Berlin/Heidelberg, Germany, 2015; pp. 215–258.

32. Fradkin, E. Application of Functional Methods in Quantum Field Theory and Quantum Statistics (II). *Nucl. Phys.* **1966**, *76*, 588–624. [CrossRef]

33. Berezin, F.; Marinov, M. Particle Spin Dynamics as the Grassmann Variant of Classical Mechanics. *Ann. Phys.* **1977**, *104*, 336–362. [CrossRef]

34. Howe, P.; Penati, S.; Pernici, M.; Townsend, P. Wave Equations for Arbitrary Spin from Quantization of the Extended Supersymmetric Spinning Particle. *Phys. Lett. B* **1988**, *215*, 555–558. [CrossRef]

35. Brink, L.; Di Vecchia, P.; Howe, P. A Lagrangian Formulation of the Classical and Quantum Dynamics of Spinning Particles. *Nucl. Phys. B* **1977**, *118*, 76–94. [CrossRef]

36. Wiegmann, P.B. Multivalued Functionals and Geometrical Approach for Quantization of Relativistic Particles and Strings. *Nucl. Phys. B* **1989**, *323*, 311–329. [CrossRef]

37. Carinena, J.F.; Garcia-Bondia, J.; Varilly, J.C. Relativistic Quantum Kinematics in the Moyal Representation. *J. Phys. A Math. Gen.* **1990**, *23*, 901–933. [CrossRef]

38. Andrzejewski, K.; Gonera, C.; Goner, J.; Kosiński, P.; Maślanka, P. Spinning Particles, Coadjoint Orbits and Hamiltonian Formalism. *arXiv* **2020**, arXiv:2008.09478.

39. Novozhilov, Y.V. *Introduction to Elementary Particle Theory*; Pergamon Press: Oxford, UK, 1975.

40. Duval, C.; Horvathy, P. Chiral Fermions as Classical Massless Spinning Particles. *Phys. Rev. D* **2015**, *91*, 045013. [CrossRef]

41. Duval, C.; Elbistan, M.; Horvathy, P.; Zhang, P.M. Wigner–Souriau Translations and Lorentz Symmetry of Chiral Fermions. *Phys. Lett. B* **2015**, *742*, 322–326. [CrossRef]

42. Andrzejewski, K.; Kijanka-Dec, A.; Kosiński, P.; Maślanka, P. Chiral Fermions, Massless Particles and Poincare Covariance. *Phys. Lett. B* **2015**, *746*, 417–423. [CrossRef]

43. Skagerstam, B. Localization of Massless Spinning Particles and the Berry Phase. *arXiv* **1992**, arXiv:hep-th/9210054.

44. Kosiński, P.; Maślanka, P. Localizability, Gauge Symmetry and Newton–Wigner Operator for Massless Particles. *Ann. Phys.* **2018**, *398*, 203–213. [CrossRef]

45. Chen, J.Y.; Son, D.T.; Stephanov, M.A.; Yee, H.U.; Yin, Y. Lorentz Invariance in Chiral Kinetic Theory. *Phys. Rev. Lett.* **2014**, *113*, 182302. [CrossRef]

46. Andrzejewski, K.; Brihaye, Y.; Gonera, C.; Gonera, J.; Kosiński, P.; Maślanka, P. The Covariance of Chiral Fermions Theory. *J. High Energy Phys.* **2019**, *8*, 11. [CrossRef]

47. Berger, L. Side-Jump Mechanism for the Hall Effect of Ferromagnets. *Phys. Rev. B* **1970**, *2*, 4559. [CrossRef]

48. Bliokh, K.Y.; Bliokh, Y.P. Topological Spin Transport of Photons: the Optical Magnus Effect and Berry Phase. *Phys. Lett. A* **2004**, *333*, 181–186. [CrossRef]

49. Onoda, M.; Murakami, S.; Nagaosa, N. Hall Effect of Light. *Phys. Rev. Lett.* **2004**, *93*, 083901. [CrossRef]

50. Duval, C.; Horvath, Z.; Horváthy, P. Geometrical Spinoptics and the Optical Hall Effect. *J. Geom. Phys.* **2007**, *57*, 925–941. [CrossRef]

51. Duval, C.; Horváth, Z.; Horváthy, P. Fermat Principle for Spinning Light. *Phys. Rev. D* **2006**, *74*, 021701. [CrossRef]

52. Bliokh, K.Y.; Nori, F. Relativistic Hall Effect. *Phys. Rev. Lett.* **2012**, *108*, 120403. [CrossRef] [PubMed]

53. Stone, M.; Dwivedi, V.; Zhou, T. Wigner Translations and the Observer Dependence of the Position of Massless Spinning Particles. *Phys. Rev. Lett.* **2015**, *114*, 210402. [CrossRef]

54. Bolonek-Lasoń, K.; Kosiński, P.; Maślanka, P. Lorentz Transformations, Sideways Shift and Massless Spinning Particles. *Phys. Lett. B* **2017**, *769*, 117–120. [CrossRef]

55. Todorov, I.T. *Conformal Description of Spinning Particles*; Springer: Berlin/Heidelberg, Germany, 1986.

56. Mack, G.; Salam, A. Finite Component Field Representations of the Conformal Group. *Ann. Phys.* **1969**, *53*, 174–202. [CrossRef]

57. Mack, G. All Unitary Ray Representations of the Conformal Group SU (2,2) with positive energy. *Commun. Math. Phys.* **1977**, *55*, 1–28. [CrossRef]

58. Gonera, J.; Kosiński, P.; Maślanka, P. Conformal Symmetry, Chiral Fermions and Semiclassical Approximation. *Phys. Lett. B* **2020**, *800*, 135111. [CrossRef]

59. Kosiński, P.; Maślanka, P. Hamiltonian Description of Conformally Invariant Elementary Systems. Work in progress.

Article

Higgs–Chern–Simons Gravity Models in $d = 2n + 1$ Dimensions

Eugen Radu [1] and D. H. Tchrakian [2,3,*]

[1] Centre for Research and Development in Mathematics and Applications (CIDMA), Department of Mathematics, University of Aveiro, Campus de Santiago, 3810-183 Aveiro, Portugal; eugen.radu@ua.pt

[2] School of Theoretical Physics, Dublin Institute for Advanced Studies, 10 Burlington Road, D04 C932 Dublin, Ireland

[3] Department of Computer Science, National University of Ireland Maynooth, Maynooth , Ireland

[*] Correspondence: tigran@stp.dias.ie

Received: 20 November 2020; Accepted: 10 December 2020; Published: 12 December 2020

Abstract: We consider a family of new Higgs–Chern–Simons (HCS) gravity models in $2n + 1$ dimensions ($n = 1, 2, 3$). This provides a generalization of the (usual) gravitational Chern–Simons (CS) gravities resulting from non-Abelian CS densities in all odd dimensions, which feature vector and scalar fields, in addition to the metric. The derivation of the new HCS gravitational (HCSG) actions follows the same method as in the *usual*-CSG case resulting from the *usual* CS densities. The HCSG result from the HCS densities, which result through a one-step descent of the Higgs–Chern–Pontryagin (HCP), with the latter being descended from Chern-Pontryagin (CP) densities in some even dimension. A preliminary study of the solutions of these models is considered, with exact solutions being reported for spacetime dimensions $d = 3, 5$.

Keywords: gravity models; Chern–Simons gravity; exact solutions

---

1. Introduction

The study of the Chern–Simons gravities (CSG) derived from non-Abelian Chern–Simons (CS) densities has started with Witten's work in Ref. [1], dealing with the $2 + 1$ dimensional case. Subsequently, Witten's results were extended to all odd dimensions by Chamseddine in Refs. [2,3]. Generic CSG models consist of superpositions of gravitational models of all possible higher order gravities in the given dimensions (leading to second order equations of motion), each appearing with a precise real numerical coefficient. These gravitational models are usually referred to as Lovelock models, which, here, we refer to as p-Einstein gravities. The integer $p \geq 0$ is the power of the Riemann curvature in the Lagrangian; the $p = 0$ term is the cosmological constant, for $p = 1$ the Ricci scalar etc.

The recent work [4,5] has proposed a new formulation of the CSG systems, which, different from the standard case in [1–3], allows for their construction in all, *both* odd and even dimensions. Following Ref. [6], let us briefly review this construction. As discussed there, the expression of the new-CS densities is found following exactly the same method as the *usual*-CS densities in odd dimensions. The *usual* CS density results from the one-step descent of the corresponding Chern–Pontryagin (CP) density. We recall that the CP density is a total-divergence

$$\Omega_{\mathrm{CP}} = \partial_i \Omega^i , \quad i = \mu, D ; \quad \mu = 1, 2, \ldots, d ; \ d = D - 1 ;$$

then the CS density is defined as the D-th component of Ω^i, namely $\Omega_{\mathrm{CS}} \stackrel{\mathrm{def.}}{=} \Omega_D$.

In the proposal put forward in [4,5], the role of the usual-CP density, which is defined in even dimensions only, is played by what we refer to as the Higgs–Chern–Pontryagin (HCP) density (see the Refs. [7,8] and in Appendix A of Ref. [9] for a discussion of HCP models). These are dimensional descendents of the *n*th CP density in $N = 2n$ dimensions, down to residual D dimensions ($D < N = 2n$). However, as a new feature, D can be either odd or even. Additionally, the relics of the gauge connection on the co-dimension(s) are Higgs scalars. The remarkable property of the HCP density $\Omega_{\mathrm{HCP}}[A, \Phi]$, which is now given in terms of both the residual gauge field A and Higgs scalar Φ, is that, like the CP density, it is also a *total divergence*

$$\Omega_{\mathrm{HCP}} = \partial_i \Omega^i , \quad i = \mu, D ; \quad \mu = 1, 2, \ldots, d ; \, d = D - 1 .$$

The corresponding new Chern–Simons density is defined by considering the one-step descent of the density Ω_i, as the D-th component of Ω_i, namely $\Omega_{\mathrm{HCS}} \overset{\mathrm{def.}}{=} \Omega_D$. In what follows, the quantity Ω_{HCS} is referred to as the Higgs–Chern–Simons (HCS) density. As mentioned above, such densities exist in both odd and even dimensions. Moreover, in any given dimension, there is an infinite family of HCS densities, following from the descent of a CP density in any dimension $N = 2n > D$. A detailed discussion of these aspects is given in Refs. [7–9]. Note that a similar definition for the HCS density was proposed in Ref. [10], but only in odd dimensions and with the Higgs scalar being a complex column, not suited to the application here.

With this definition of the HCS densities, the construction of the corresponding gravitational theories is done in the same spirit as in [1–3]. In any given dimension, there is an infinite family of such theories, each that result from the infinite family of HCS densities. Working in $d = D - 1$ dimensions, the gauge group is chosen to be $SO(d)$, while the Higgs multiplet is chosen to be a D-component *isovector* of $SO(D)$ (These choices coincide with the representations that yield monopoles on $\mathbf{R}^d$, as described in [7]). The central point in the construction of both CS and HCS gravity models is the identification of the non-Abelian (nA) $SO(D)$ connection in $d = D - 1$ dimensions (Note that no choice for the signature of the space is make at this stage), with the spin-connection ω_μ^{ab} and the *Vielbein* e_μ^a, ($\mu = 1, 2, 3$; a=1,2,3). Following the prescription presented in [1–3], we define

$$A_\mu \;\; = \;\; -\frac{1}{2}\, \omega_\mu^{ab}\, \gamma_{ab} + \kappa\, e_\mu^a\, \gamma_{aD} \;\; \Rightarrow \;\; F_{\mu\nu} = -\frac{1}{2}\left(R_{\mu\nu}^{ab} - \kappa^2\, e_{[\mu}^a\, e_{\nu]}^b\right) \gamma_{ab} , \tag{1}$$

$(\gamma^{ab}, \gamma^{aD})$ being the Dirac gamma matrices that are used in the representation of the algebra of $SO(D)$. Note that the presence of the constant κ in the above expression (with dimensions L^{-1}), compensating the difference of the dimensions of the spin-connection and the *Dreibein*. In (1),

$$R_{\mu\nu}^{ab} = \partial_{[\mu}\omega_{\nu]}^{ab} + (\omega_{[\mu}\omega_{\nu]})^{ab}$$

is the Riemann curvature.

In the HCS case, in addition to (1), we supplement (1) with the prescription for the Higgs scalar Φ,

$$2^{-1}\Phi = (\phi^a\, \gamma_{a,D+1} + \psi\, \gamma_{D,D+1}) \;\; \Rightarrow \;\; 2^{-1}D_\mu\Phi = (D_\mu\phi^a - \kappa\, e_\mu^a\, \psi)\gamma_{a,D+1} + (\partial_\mu\psi + \kappa\, e_\mu^a\, \phi^a)\gamma_{D,D+1} \tag{2}$$

which clearly displays the *iso-D-vector* (ϕ^a, ϕ^D), which is split into the D *component* frame-vector field ϕ^a and the scalar field $\psi \equiv \phi^D$. Additionally, we define the covariant derivative $D_\mu\Phi$ of the Higgs scalar

$$D_\mu\phi^a = \partial_\mu\phi^a + \omega_\mu^{ab}\phi^b . \tag{3}$$

Additionally, note that ϕ^a is a vector field (with $\phi_\mu = e_\mu^a\phi^a$ in a coordinate frame), which, however, has rather unusual dynamics, as will be seen below. As such, ϕ^a is not a gauge (masless or massive) field; it has rather a geometric content.

In fact, one remarks that the fields (ϕ^a, ψ) are not usual matter fields, like gauge fields or Higgs scalar in the Standard Model of particle physics. In the latter cases, the covariant derivatives are not defined by the (gravitational) spin-connection, while, here, they are, as seen in (2). Thus, in this sense, (ϕ^a, ψ) are like spinor fields (although their action look different). As such, theories like the one that is proposed here can support solutions with torsion, a possibility, which, however, is not explored in this work. However, the analogy with spinors is incomplete, since the fields (ϕ^a, ψ) are rather 'gravitational coordinates', as they originate from the Higgs field Φ of the nA gauge theory. Thus, as seen from (2), (ϕ^a, ψ) are on the same footing as the *usual* 'gravitational coordinates' $(\omega_\mu^{ab}, e_\mu^a)$. In fact, this provides the main physical motivation for their study, since such models can be seen as extensions of the usual CSG's (reducing to them in the limit of vanishing (ϕ^a, ψ)). Therefore, it is interesting to see what are the new features introduced by extending the CSG's to allow for nonzero (ϕ^a, ψ). Moreover, finding how the BTZ-like CSG's black hole solutions in Ref. [11–13] are deformed by the (ϕ^a, ψ)-fields is an interesting mathematical problem in itself.

The gravitational models resulting from the Higgs–CS (HCS) densities via (1) and (2) are referred to as HCS gravities [4] (HCSG). In this report, we restrict our study to the simplest HCSG models in $2n - 1$ dimensions, namely to the HCSG models that result from the HCS density descended from the HCP densities in 6, 8, and 10 dimensions. These models are extensions of the *usual* Chern–Simons gravities [1–3], possessing an additional sector in terms of (ϕ^a, ψ).

This paper is organized, as follows. In the next Section, we present the explicit form of the resulting HCSG Langrangians for $d = 3, 5, 7$, with the connection with the usual CSG also being reviewed. A preliminary investigation of the simplest solutions in $d = 3, 5$ dimensions is considered in Section 3, extending the study (for the $n = 1$ case) in Ref. [6]. Section 4 presents the concluding remarks.

2. HCSG Models in $2n + 1$ Dimensions, $n = 1, 2, 3$

2.1. General Expressions

The Higgs–Chern–Simons densities (HCS) considered here are the "simplest" examples in $2 + 1$, $4 + 1$, $6 + 1$ dimensions. Simply, we mean that the Higgs–Chern–Simons (HCS) density that is employed to construct the HCS gravity (HCSG) is the one resulting from the "simplest" Higgs–Chern–Pontryagin (HCP) density, which is defined in one dimension higher, namely in *four*, *six*, and *eight* dimensions, respectively. Now, in *any* dimension, HCP densities can be constructed as dimensional descendants of a CP density in $2n > 4$ dimensions; hence, it is reasonable to describe the "simplest" cases here to be the HCP densities in 4, 6, 8 dimensions, which descend from the CP densities in $2n = 6$, 8, 10 dimensions, respectively, i.e., those descended by the minimal (nontrivial) number of dimensions, namely by *two* dimensions.

Because, like the CP density, the HCP density is a *total divergence*, then the corresponding HCS density results from the usual *one-step* descent, which, in the cases at hand, are those from 4, 6, 8 to 3, 5, 7 dimensions, with the corresponding expressions

$$\Omega_{\text{HCS}}^{(3,6)} = \eta^2 \Omega_{\text{CS}}^{(3)} + \varepsilon^{\mu\nu\lambda} \text{Tr}\, \gamma_5 D_\lambda \Phi (\Phi F_{\mu\nu} + F_{\mu\nu} \Phi), \tag{4}$$

$$\Omega_{\text{HCS}}^{(5,8)} = \eta^2 \Omega_{\text{CS}}^{(5)} + \varepsilon^{\mu\nu\rho\sigma\lambda} \text{Tr}\, \gamma_7 D_\lambda \Phi \left(\Phi F_{\mu\nu} F_{\rho\sigma} + F_{\mu\nu} \Phi F_{\rho\sigma} + F_{\mu\nu} F_{\rho\sigma} \Phi \right), \tag{5}$$

$$\Omega_{\text{HCS}}^{(7,10)} = \eta^2 \Omega_{\text{CS}}^{(7)} + \varepsilon^{\mu\nu\rho\sigma\tau\lambda\kappa} \text{Tr}\, \gamma_9 D_\kappa \Phi (\Phi F_{\mu\nu} F_{\rho\sigma} F_{\tau\lambda} + F_{\mu\nu} \Phi F_{\rho\sigma} F_{\tau\lambda}$$
$$+ F_{\mu\nu} F_{\rho\sigma} \Phi F_{\tau\lambda} + F_{\mu\nu} F_{\rho\sigma} F_{\tau\lambda} \Phi). \tag{6}$$

Let us remark that the HCS densities (4)–(6) are the "simplest" examples in these dimensions, which arise from the descents of the Chern–Pontryagin densities in 6, 8, and 10 dimensions, respectively. It may also be interesting to display HCS densities that arise from CP densities in higher dimensions. To this end, we consider the HCS density in 3 dimensions that result from the descent from the CP in 8 dimensions

$$\Omega_{\text{HCS}}^{(3,8)} = 2\eta^4 \Omega_{\text{CS}}^{(3)} + (2\eta^2 - |\phi^a|^2 - \phi^2)\, \varepsilon^{\mu\nu\lambda} \text{Tr}\, \gamma_5\, D_\lambda \Phi (\Phi F_{\mu\nu} + F_{\mu\nu} \Phi) \tag{7}$$

The leading term $\Omega_{CS}^{(3)}$ in (4) and (7), and the leading terms $\Omega_{CS}^{(5)}$ and $\Omega_{CS}^{(7)}$ in (5) and (6) are the usual CS densities for the $SO(D)$, $D = 4, 6, 8$ gauge connection,

$$\Omega_{CS}^{(3)} = \varepsilon^{\lambda\mu\nu}\mathrm{Tr}\,\gamma_5 A_\lambda \left[F_{\mu\nu} - \frac{2}{3} A_\mu A_\nu \right], \tag{8}$$

$$\Omega_{CS}^{(5)} = \varepsilon^{\lambda\mu\nu\rho\sigma}\mathrm{Tr}\,\gamma_7 A_\lambda \left[F_{\mu\nu} F_{\rho\sigma} - F_{\mu\nu} A_\rho A_\sigma + \frac{2}{5} A_\mu A_\nu A_\rho A_\sigma \right], \tag{9}$$

$$\Omega_{CS}^{(7)} = \varepsilon^{\lambda\mu\nu\rho\sigma\tau\kappa}\mathrm{Tr}\,\gamma_9 A_\lambda \left[F_{\mu\nu} F_{\rho\sigma} F_{\tau\kappa} - \frac{4}{5} F_{\mu\nu} F_{\rho\sigma} A_\tau A_\kappa - \frac{2}{5} F_{\mu\nu} A_\rho A_\sigma F_{\tau\kappa} \right.$$
$$\left. + \frac{4}{5} F_{\mu\nu} A_\rho A_\sigma A_\tau A_\kappa - \frac{8}{35} A_\mu A_\nu A_\rho A_\sigma A_\tau A_\kappa \right]. \tag{10}$$

In (4)–(6), the Higgs scalar Φ, the gauge connection, and the constant η have the dimensions of L^{-1}.

Applying the prescriptions (1) and (2) to (4)–(7) yields the required HCS gravitational (HCSG) models. In order to express these compactly, we adopt the abbreviated notation

$$\bar{R}_{\mu\nu}^{ab} = R_{\mu\nu}^{ab} - \kappa^2 e_{[\mu}^a e_{\nu]}^b, \tag{11}$$

together with

$$\phi_\mu^a = D_\mu \phi^a - \kappa e_\mu^a \psi, \tag{12}$$
$$\psi_\mu = \partial_\mu \psi + \kappa e_\mu^a \phi^a, \tag{13}$$

where $R_{\mu\nu}^{ab}$ is the Riemann curvature and $D_\mu \phi^a$ is the covariant derivative (3), of the frame-vector field ϕ^a.

In terms of which the HCSG Lagrangians in $d = 3, 5,$ and 7 dimensions, which result from the HCS densities (4)–(6), are

$$\mathcal{L}_{HCSG}^{(3)} = \eta^2 \kappa\, \mathcal{L}_{CSG}^{(3)} + \varepsilon^{\mu\nu\lambda} \varepsilon_{abc} \bar{R}_{\mu\nu}^{ab} (\psi\,\phi_\lambda^c - \phi^c \psi_\lambda), \tag{14}$$

$$\mathcal{L}_{HCSG}^{(5)} = \eta^2 \kappa\, \mathcal{L}_{CSG}^{(5)} - \frac{3}{4} \varepsilon^{\mu\nu\rho\sigma\lambda} \varepsilon_{abcde} \bar{R}_{\mu\nu}^{ab} \bar{R}_{\rho\sigma}^{cd} (\psi\,\phi_\lambda^c - \phi^c \psi_\lambda), \tag{15}$$

$$\mathcal{L}_{HCSG}^{(7)} = \eta^2 \kappa\, \mathcal{L}_{CSG}^{(7)} + 2\varepsilon^{\mu\nu\rho\sigma\tau\lambda\kappa} \varepsilon_{abcdefg} \bar{R}_{\mu\nu}^{ab} \bar{R}_{\rho\sigma}^{cd} \bar{R}_{\tau\lambda}^{ef} (\psi\,\phi_\lambda^c - \phi^c \psi_\lambda), \tag{16}$$

while, the gravitational model that arises from (7) is

$$\mathcal{L}_{HCSG}^{(3,8)} = 2\eta^4 \kappa\, \mathcal{L}_{CSG}^{(3)} + \varepsilon^{\mu\nu\lambda} \varepsilon_{abc} (2\eta^2 - |\phi^d|^2 - \psi^2)\, \bar{R}_{\mu\nu}^{ab} (\psi\,\phi_\lambda^c - \phi^c \psi_\lambda). \tag{17}$$

In (14)–(16), $\mathcal{L}_{CSG}^{(3)}$, $\mathcal{L}_{CSG}^{(5)}$ and $\mathcal{L}_{CSG}^{(7)}$ are the usual Chern–Simons gravities (CSG)

$$\mathcal{L}_{CSG}^{(3)} = -\varepsilon^{\mu\nu\lambda} \varepsilon_{abc} \left(R_{\mu\nu}^{ab} - \frac{2}{3} \kappa^2 e_\mu^a e_\nu^b \right) e_\lambda^c, \tag{18}$$

$$\mathcal{L}_{CSG}^{(5)} = \varepsilon^{\mu\nu\rho\sigma\lambda} \varepsilon_{abcde} \left(\frac{3}{4} R_{\mu\nu}^{ab} R_{\rho\sigma}^{cd} - \kappa^2 R_{\mu\nu}^{ab} e_\rho^c e_\sigma^d + \frac{3}{5} \kappa^4 e_\mu^a e_\nu^b e_\rho^c e_\sigma^d \right) e_\lambda^e, \tag{19}$$

$$\mathcal{L}_{CSG}^{(7)} = -\varepsilon^{\mu\nu\rho\sigma\tau\kappa\lambda} \varepsilon_{abcdefg} \left(\frac{1}{8} R_{\mu\nu}^{ab} R_{\rho\sigma}^{cd} R_{\tau\kappa}^{ef} - \frac{1}{4} \kappa^2 R_{\mu\nu}^{ab} R_{\rho\sigma}^{cd} e_\tau^e e_\kappa^f \right.$$
$$\left. + \frac{3}{10} \kappa^4 R_{\mu\nu}^{ab} e_\rho^c e_\sigma^d e_\tau^e e_\kappa^f - \frac{1}{7} \kappa^6 e_\mu^a e_\nu^b e_\rho^c e_\sigma^d e_\tau^e e_\kappa^f \right) e_\lambda^g. \tag{20}$$

It is easy to express the HCSG Lagranian for all n by the extrapolation of (14)–(16).

Given the models (14)–(16) , the corresponding equations of motion are found by taking the variation of the action w.r.t., the vielbein e^a_λ, together with (ϕ^a, ψ). However, these equations have a simple enough expression for $d = 3$ only, see Ref. [6].

2.2. The General CSG Lagrangians and the Connection with the Einstein–Lovelock Hierarchy

Following the previous study [6], we consider a spacetime with Minkowskian signature, and replace

$$\kappa \to i\kappa, \quad h \to -ih. \tag{21}$$

in the Lagrangians (14)–(16). With this choice, setting $\phi = \phi^a = 0$ results in (pure gravity) CS Lagrangian in $d = 2n + 1$ dimensions, with a negative cosmological constant Λ. Additionally, note that one can $\eta = 1$ without any loss of generality, a choice that we employ for the rest of this work.

The CS Lagrangian in $d = 2n + 1$ dimensions can be viewed as a particular case of a generic model that consists in a superposition of all allowed Einstein–Lovelock terms in that dimension, with

$$\mathcal{L}^{(2n+1)}_{\text{CSG}} = \sum_{p=0}^{n} \alpha_{(p)} L_{(p)}, \tag{22}$$

with the following definition for the $p-$th term in the Lovelock hierarchy.

$$L_{(p)} = \frac{p!}{2^p} \delta^{\mu_1}_{[\rho_1} \cdots \delta^{\mu_p}_{\rho_p]} R_{\mu_1 \nu_1}{}^{\rho_1 \sigma_1} \cdots R_{\mu_p \nu_p}{}^{\rho_p \sigma_p}. \tag{23}$$

The normalization of each term in (23) has been chosen in order to make contact with the usual conventions in the literature on Lovelock gravities solutions. As such, $L_{(p)} = R^p + \dots$, the first terms (up to $d = 7$) being

$$L_{(0)} = 1, \quad L_{(1)} = R, \quad L_{(2)} = R^2 - 4R_{\mu\nu}R^{\mu\nu} + R_{\mu\nu\rho\sigma}R^{\mu\nu\rho\sigma}, \tag{24}$$

$$L_{(3)} = R^3 - 12RR_{\mu\nu}R^{\mu\nu} + 16R_{\mu\nu}R^\mu{}_\rho R^{\nu\rho} + 24R_{\mu\nu}R_{\rho\sigma}R^{\mu\rho\nu\sigma} + 3RR_{\mu\nu\rho\sigma}R^{\mu\nu\rho\sigma}$$
$$- 24R_{\mu\nu}R^\mu{}_{\rho\sigma\kappa}R^{\nu\rho\sigma\kappa} + 4R_{\mu\nu\rho\sigma}R^{\mu\nu\eta\zeta}R^{\rho\sigma}{}_{\eta\zeta} - 8R_{\mu\rho\nu\sigma}R^\mu{}_\eta{}^\nu{}_\zeta R^{\rho\eta\sigma\zeta}. \tag{25}$$

Additionally, to make contact with the usual GR conventions, we take

$$\alpha_{(0)} = -2\Lambda, \quad \alpha_{(1)} = 1. \tag{26}$$

In general, the coefficients $\alpha_{(p)}$ are arbitrary. However, they are fixed in a CGS model, with the general expression

$$\alpha_{(p)} = (-1)^{p+1} \frac{1}{\Lambda^{p-1}} \frac{(d-2)^p}{2^{p-1}p!(d-2p)} \frac{(d-2p)!!}{(d-2)!!}, \tag{27}$$

where $\Lambda = -2(d-2)\kappa^2$.

3. The Solutions

Given the above models (14)–(16), it is interesting to inquire which are the simplest solutions with nonvanishing fields (ϕ^a, ψ). In what follows, we study this question for the first two dimensions $d = 3, 5$, and then contrast the results.

3.1. The $d = 3$ Case

We consider a static line element with

$$ds^2 = \frac{dr^2}{N(r)} + r^2 d\varphi^2 - N(r)e^{-2\delta(r)}dt^2 \tag{28}$$

where r, t are the radial and time coordinates, respectively, while φ is the azimuthal coordinate. Working in a coordinate basis, a consistent Ansaz for the fields (ϕ^a, ψ) reads

$$\phi = f(r)dr + g(r)dt, \quad \psi \equiv h(r). \tag{29}$$

Subsequently, a straightforward computation leads to the following exact solution of the full set of equations of motion:

$$N(r) = \kappa^2 r^2 - n_0, \quad \delta(r) = 0, \tag{30}$$

$$f(r) = \frac{c_0}{\kappa} + \frac{c_1}{N}, \quad g(r) = \sqrt{c_1^2 + c_2 N(r)}, \quad \psi \equiv h(r) = c_0 r, \tag{31}$$

with n_0, c_0, c_1, c_2 arbitrary constants. This provides a generalization of the solution that was reported in [6], which is recovered (note that the solution presented in [6] is expressed in a *dreibein* basis with $e_r = 1/\sqrt{N}$, $e_\varphi = r d\varphi$, $e_t = \sqrt{N}dt$), for $c_1 = c_2 = 0$.

One can see that the choice $n_0 = -1$ corresponds to a globally AdS$_3$ geometry, while, for $n_0 > 0$, the BTZ black hole (BH) geometry [14] is recovered. In both cases, the fields (ϕ^a, ψ) do not backreact on the spacetime (thus, their contribution to the *r.h.s.* of the Einstein equations with negative cosmological constant vanishes (the analogy of these solutions with self-dual Yang–Mills instantons in a curved space geometry [15–17], was noticed in Ref. [6]. There, we dubbed these closed form solutions as effectively vacuum confiurations. Another interesting analogy is provided by the 'stealth' BH solutions in various alternative models of gravity (see e.g., [18,19] and references therein). Such configurations feature a nontrivial scalar field, while the geometry is still that of the (vacuum) general relativity solutions)). However, (ϕ^a, ψ) possess a nonstandard behaviour (e.g., both ψ and $|\vec{\phi}|^2$ diverge as $r \to \infty$). Moreover, Ref. [6] has given arguments that (at least in the $c_1 = c_2$ limit), solutions (30) and (31) appear to be unique. Because the discussion here in $d = 3$ is similar to that in the $d = 5$ case, we restrict to the discussion of the latter, as below.

Finally, we mention the existence of a generalization of the solution in Ref. [6] for a spinning BTZ background. The line element in this case reads

$$ds^2 = \frac{dr^2}{N(r)} + r^2(d\varphi - W(r)dt)^2 - N(r)dt^2, \quad \text{where } N(r) = \kappa^2 r^2 - n_0 + \frac{J^2}{r^2}, \quad W(r) = \frac{J}{r^2}, \tag{32}$$

with J representing the angular momentum. While the expression of the scalar field remains the same, the function is more complicated,

$$h(r) = c_0 r, \quad f(r) = \frac{c_0}{\kappa}\left(1 + \frac{J}{r\sqrt{N}}\right). \tag{33}$$

One can easily see that all of the unusual features noticed in the static case for functions $f\ h$ are present also in this case.

3.2. The $d = 5$ Case

3.2.1. An Exact Solution

The metric Ansatz here is more complicated, with the S^1 direction in the $d = 3$ line-element (28) being replaced with a surface of constant curvature. As such, we consider a general line-element

$$ds^2 = \frac{dr^2}{N(r)} + r^2 d\Sigma_{k,3}^2 - N(r)e^{-2\delta(r)} N(r)dt^2 \tag{34}$$

with $k = 0, \pm 1$, while the three-dimensional metric $d\Sigma^2_{k,3}$ is

$$d\Sigma^2_{k,3} = \begin{cases} d\Omega^2_3 & \text{for } k = +1 \\ \sum^3_{i=1} dx^2_i & \text{for } k = 0 \\ d\Xi^2_3 & \text{for } k = -1 \,. \end{cases} \tag{35}$$

In the above relation, $d\Omega^2_3$ is the unit metric on S^3 ($k = 1$); for $k = 0$, we have a flat three-surface; while, for $k = -1$, one considers a three-dimensional hyperbolic space, whose unit metric $d\Xi^2_3$ can be obtained by analytic continuation of that on S^3.

The ansatz for the fields (ϕ^a, ψ) is still given by (29). Within this framework, the following closed-form solution of the field equations has been found

$$N(r) = \kappa^2 r^2 + k, \quad \delta(r) = 0, \tag{36}$$

together with

$$h(r) = \frac{c_0}{r}, \quad f(r) = \frac{c_0}{\kappa r^2}\left(-1 + (k^2 + k - 1)\sqrt{1 + \frac{3\kappa^2 r^2}{N(r)}}\right), \quad g(r) = 0\,, \tag{37}$$

with c_0 being an arbitrary constant. The corresponding line elements, as implied by (36), are well known, which correspond to three different parametrization of AdS_5 spacetime. Although they possess the same (maximal) number of Killing symmetries, they present different global properties, the case $k = 0$ corresponding to a Poincaré patch and the globally AdS_5 spacetime being found for $k = 1$ (see e.g., the discussion in Ref. [20]).

As with the $d = 3$ case presented above, the fields (ϕ^a, ψ) do not backreact on the spacetime geometry, i.e., their effective energy-momentum tensor vanishes again. However, while, this time, both ψ and $|\vec{\phi}|^2$ are finite as $r \to \infty$, they diverge at the minimal value of r (which is $r = 0$ for $k = 0, 1$ and $r = 1/\kappa$ for $k = -1$).

Finally, let us remark that the expressions (37) for (ϕ^a, ψ) are also compatible with a different background, as described by the line-element (where $k = 0, \pm 1$)

$$ds^2 = \frac{dr^2}{N(r)} + r^2 d\Sigma^2_{k,2} + r^2 dz^2 - N(r)dt^2, \quad \text{with } N(r) = \kappa^2 r^2 + \frac{k}{3}. \tag{38}$$

where $-\infty < z < \infty$, while $d\Sigma^2_{k,2}$ the metric on a two-dimensional surface of constant $2k$-curvature. For $k = 0$ the Poincaré patch of AdS_5 spacetime is recovered; the case $k = 1$ corresponds to a vortex-type geometry, while a black string is recovered for $k = -1$. As with the line-element (34), this is also a solution of the equations of motion for $f = g = h = 0$, whose existence is noted in Ref. [21].

3.2.2. No Backreacting Solutions on a Fixed Black Hole Background

For $\phi^a = \psi = 0$, the CSG equations of motion (with the line-element (34)) possess the exact solution

$$N(r) = \kappa^2 r^2 - n_0, \quad \delta(r) = 0, \tag{39}$$

with n_0 being an arbitrary constant. The AdS_5 line-element that is discussed above is a particular case here, corresponding to the choice $n_0 = -k$. However, $n_0 > 0$ leads to a BTZ-like BH geometry [11–13], with an horizon being located at $\sqrt{n_0}/\kappa$. Moreover, one can show that the same expression of the metric functions $N(r), \delta(r)$ solves the CS gravity equation in all $d = 2n + 1$ dimensions (for a choice of the line element that is similar to (34), $d\Sigma^2_{k,3}$ being replaced with its higher dimensional generalization); see the discussion in the recent work [22] and the references there.

A major difference between the $d = 3$ and $d = 5$ cases appear to be that, for $d = 5$, we could not extend (36) to include the case of a BH solution, $n_0 \neq k$. However, this result follows directly from the structure of the field equations, which is different for $d = 3$ and $d = 5$. Let us assume that the geometry (34) with (N, δ), as given by (39)), is a solution of the $d = 5$ model. Subsequently, the equations for the functions f, g, h take the simple form (note that the relations below are for $k = 1$, only; however, a similar result is also found for $k = 0, -1$):

$$
\begin{aligned}
&h' - \kappa f - \frac{\kappa}{2}\left(1 - \frac{1}{N^2}\right)g = 0, \\
&h' - \frac{1}{2}\kappa(1 - N^2)f - \frac{n_0 + \kappa^2 r^2 N^2}{2rN}h - \kappa g = 0, \\
&g' - N^2 f' + \frac{\kappa^2 r^2 N^2 - 2N - 3n_0}{2rN}g - \frac{2n_0 + 3N}{r}f + 3\kappa N h = 0.
\end{aligned}
\tag{40}
$$

It directly follows that both f and g can be expressed in terms of h, with

$$
f = \frac{1}{2}\left(-1 + \frac{1}{N^2}\right)g + \frac{h'}{\kappa},
\tag{41}
$$

and

$$
g = \frac{2N^2}{\kappa(1 + N^2)}h' - \frac{2N(n_0 + \kappa^2 r^2 N^2)}{\kappa r(1 + N^2)^2}h = 0 \, .
\tag{42}
$$

The scalar h is a solution of an equation of the form

$$
(1 + n_0)hU(r) = 0, \quad \text{with} \quad U(r) = \sum_{k=0}^{5} c_k(\kappa, n_0)r^{2k},
\tag{43}
$$

with the explicit form of c_k being irrelevant. As such, for $h \neq 0$, the only choice is $n_0 = -1$, i.e., a globally AdS$_5$ spacetime. Subsequently, the matter fields equations are satisfied, the functions f and g being fixed by h. The expression of the scalar h is found by imposing the gravity equations that are to be satisfied for the above choice of the geometry, which result in the solution (37).

A similar computation for $d = 3$ leads again to a set of three equations for (ϕ^a, ψ), which, again, reduce to a single equation for $h(r)$. However, this time, this equation is multiplied with a factor $N' - 2\kappa^2 r$. Therefore, the choice $N = \kappa^2 r^2 - n_0$, with n_0 arbitrary, is now allowed. The solution (31) is recovered when imposing the Einstein equation also to be satisfied.

Returning to the $d = 5$ case, one may ask whether a more general solution exists, with the functions (ϕ^a, ψ) backreacting on the spacetime metric and being finite everywhere. The answer seems to be negative, although we do not have a definite proof. An indication comes from our attempt to construct a numerical solution. Here, one starts by noticing that, starting with the general framework (29) and (34), the function f can be eliminated (as found from the field equations), with

$$
f(r) = \frac{h'(r)}{\kappa},
\tag{44}
$$

where we assume that globally AdS spacetime is *not* a solution. Additionally, one can prove that $g = 0$ is a consistent truncation of the model. As such, we are left with three ordinary differential equations for the functions h, N, and δ. Restricting to the most interesting $k = 1$ case (i.e., a globally AdS$_5$ background), we have attempted to construct deformations of the line element (36), with a regular origin and usual AdS$_5$ asymptotics, which would represent particle-like solitonic configurations. In our approach, we assume that the small-r solution possesses a power series expansion, with

$$
N(r) = \sum_{k \geq 0} n_{(k)}r^k, \quad \delta(r) = \sum_{k \geq 0} \delta_{(k)}r^k \quad \text{and} \quad h(r) = \sum_{k \geq 0} h_{(k)}r^k,
\tag{45}
$$

where $n_{(k)}$, $\delta_{(k)}$, and $h_{(k)}$ are real numbers (and $n_0 = 1$) subject to a tower of algebraic conditions, as implied by the field equations. Starting with the above small-r expansion, we have integrated the HCS equations of motion, searching for solutions with $N(r) \to \kappa^2 r^2 + const.$, $\delta \to 0$ and $h(r) \to h_0$ (with h_0 a constant), as $r \to \infty$. However, we have failed to find any numerical indication for the existence of such configurations, the solutions that possess a pathological behaviour for any considered set of initial conditions at $r = 0$, typically with the occurence of a divergence at some finite r.

A similar result also holds for BH configurations, in which case we assume the existence of an horizon at some $r = r_H > 0$, with $N(r_H) = 0$, while $\delta(r_H)$ and $h(r_H)$ are nonzero and finite.

Finally, let us remark that, although a definite proof is missing, the above (numerical) results follow the spirit of the 'no hair' theorems [23–25], as expressed in the conjecture that there are no BH solutions with matter fields that do not possess (asymptotically) measured quantities that are subject to a Gauss Law.

4. Conclusions

Chern–Simons gravity (CSG) models in $d = 2n + 1$ dimensions were extensively studied in the literature, starting with Witten's work for $d = 3$ [1], where the gravitational model is described by the Einstein–Hilbert Lagrangian with a cosmological constant. In the $d > 3$ case, such systems consist of specific superpositions of gravitational Lagrangians featuring all possible powers of the Riemann curvature of the given dimension, each appearing with a precise numerical coefficient. The main purpose of this paper was to propose a generalization of the CSG model, with a Lagrangian, which, in addition to the (standard) CSG Lagrangian, features new terms that are described by a frame-vector field ϕ^a and a scalar field ψ. Like the CSG, which result from the non-Abelian (nA) Chern–Simons (CS) densities, these new Lagrangians result from a new class of CS densities, which, in addition to the nA gauge field, feature an algebra-valued Higgs scalar. Like the usual nA CS densities, which result from the usual Chern–Pontryagin (CP) densities, these new CS densities are constructed in the same way, but now from the dimensional descendents of the CP densities that feature the Higgs scalar. The latter are referred to as Higgs–Chern–Pontryagin (HCS) [7–9] densities, and they are the building blocks for the generalised CSG's, namely the HCSG's [4,5] studied here.

It should be noted at this stage that the construction of HCSG's is not only confined to odd dimensions, since the HCS from which they are constructed are defined is both odd *and* even dimensions. The main reason that we have restricted our attention to odd dimensions in these preliminary investigations is that only in odd dimensions there exist CSG's, which can provide a background for the new gravitational field configurations. In even dimensional spacetimes, the HCSG models, as typified by the $3 + 1$ dimensional examples in Refs. [4,5], also consist of frame-vector and scalar fields (ϕ^a, ψ) that interact with the gravitational *Vielbein* e_μ^a (or the metric). These Lagrangians are invariant under gravitational gauge transformations; however, different from the odd dimensional case in this work, they do not feature (gauge-variant, pure gravity) CSG terms, with their action mixing the contribution of (ϕ^a, ψ), e_μ^a fields.

The new fields (ϕ^a, ψ) display non-standard dynamics, in that they feature linear 'velocity coordinates', rather than the standard 'velocity squared' kinetic terms. It may be relevant to stress that (ϕ^a, ψ) can be seen as 'gravitational coordinates', rather than usual matter fields, since, on the level of the HCS densities from which the HCSG result, the Higgs scalar is on the same footing as the non Abelian gauge connection.

The present work, which is a continuation of that done in Ref. [6] for the the lowest dimension $d = 3$, provides the explicit expression of the HCS Lagrangians up to $d = 7$, together with an investigation of the simplest solutions for $d = 3, 5$. These solutions have the property that they do not backreact on the spacetime geometry in common, i.e., their effective energy-momentum tensor vanishes. However, while, for $d = 3$, this includes the case of BTZ BH, for $d = 5$ only a maximally symmetric AdS background is allowed. We attribute this feature to the fact that the BTZ BH possesses the same amount of symmetries as pure AdS_3, being a global identification of it [14,26]. On the other hand, the case of $d > 3$ BHs in CSG are different; although their line-element is still BTZ-like [22], they are less symmetric than the AdS$_d$ background.

Finally, for $d = 3$, the Ref. [6] has provided (numerical) evidence for the existence of BTZ-like BH also with standard asymptotics for the fields (ϕ^a, ψ), provided that the action is supplemented with a Maxwell field. We conjecture that a similar property holds in the higher dimensional case. In this respect, it may be interesting to consider the HCSG systems in the presence of non-Abelian matter (in $d > 3$), or Skyrme scalars, in order to search for regular solutions.

Author Contributions: The two authors contributed equally to the conceptualization, methodology, software, validation, formal analysis, investigation, resources, data curation, writing—original draft preparation, writing—review and editing, visualization, supervision, project administration, funding acquisition. All authors have read and agreed to the published version of the manuscript.

Funding: The work of E.R. is supported by the Center for Research and Development in Mathematics and Applications (CIDMA) through the Portuguese Foundation for Science and Technology (FCT-Fundacao para a Ciência e a Tecnologia), references UIDB/04106/2020 and UIDP/04106/2020, and by national funds (OE), through FCT, I.P., in the scope of the framework contract foreseen in the numbers 4, 5 and 6 of the article 23,of the Decree-Law 57/2016, of August 29, changed by Law 57/2017, of July 19. We acknowledge support from the projects PTDC/FIS-OUT/28407/2017 and CERN/FIS-PAR/0027/2019. This work has further been supported by the European Union's Horizon 2020 research and innovation (RISE) programme H2020-MSCA-RISE-2017 Grant No. FunFiCO-777740. The authors would like to acknowledge networking support by the COST Action CA16104.

Acknowledgments: We are grateful to Ruben Manvelyan for useful discussions.

Conflicts of Interest: The authors declare no conflict of interest.

References

1. Witten, E. (2+1)-dimensional gravity as an exactly soluble system. *Nucl. Phys. B* **1988**, *311*, 46–78.
2. Chamseddine, A.H. Topological gauge theory of gravity in five-dimensions and all odd dimensions. *Phys. Lett. B* **1989**, *233*, 291.
3. Chamseddine, A.H. Topological gravity and supergravity in various dimensions. *Nucl. Phys. B* **1990**, *346*, 213.
4. Tchrakian, D.H. Chern-Simons gravities (CSG) and gravitational Chern-Simons (GCS) densities in all dimensions. *Phys. At. Nucl.* **2018**, *81*, 930–938.
5. Radu, E.; Tchrakian, D.H. Gravitational Chern–Simons, and Chern–Simons Gravity in All Dimensions. *Phys. Part. Nucl. Lett.* **2020**, *17*, 753–759.
6. Radu, E.; Tchrakian, D. A Higgs–Chern–Simons gravity model in 2+ 1 dimensions. *Class. Quant. Grav.* **2018**, *35*, 175012.
7. Tchrakian, D.H. Notes on Yang–Mills–Higgs monopoles and dyons on, and Chern–Simons–Higgs solitons on: Dimensional reduction of Chern–Pontryagin densities. *J. Phys. A* **2011**, *44*, 343001.
8. Radu, E.; Tchrakian, D.H. New Chern-Simons densities in both odd and even dimensions. In *Low Dimensional Physics and Gauge Principles: Matinyan's Festschrift*; World Scientific: Singapore, 2013; pp. 227–240.
9. Tchrakian, D.H. Higgs-and Skyrme–Chern–Simons densities in all dimensions. *J. Phys. A* **2015**, *48*, 375401.
10. Szabo, R.J.; Valdivia, O. Covariant quiver gauge theories. *J. High Energy Phys.* **2014**, *2014*, 144.
11. Cai, R.G.; Soh, K.S. Topological black holes in the dimensionally continued gravity. *Phys. Rev. D* **1999**, *59*, 044013.
12. Crisostomo, J.; Troncoso, R.; Zanelli, J. Black hole scan. *Phys. Rev. D* **2000**, *62*, 084013.
13. Aiello, M.; Ferraro, R.; Giribet, G. Exact solutions of Lovelock-Born-Infeld black holes. *Phys. Rev. D* **2004**, *70*, 104014.
14. Banados, M.; Teitelboim, C.; Zanelli, J. Black hole in three-dimensional spacetime. *Phys. Rev. Lett.* **1992**, *69*, 1849.
15. Charap, J.M.; Duff, M.J. Gravitational effects on Yang-Mills topology. *Phys. Lett. B* **1977B**, *69*, 445–447.
16. Brihaye, Y.; Radu, E. On d= 4 Yang-Mills instantons in a spherically symmetric background. *Europhys. Lett.* **2006**, *75*, 730.
17. Radu, E.; Tchrakian, D.H.; Yang, Y. Spherically symmetric self-dual Yang-Mills instantons on curved backgrounds in all even dimensions *Phys. Rev. D* **2008**, *77*, 044017.

18. Bernardo, R.C.; Celestial, J.; Vega, I. Stealth black holes in shift symmetric kinetic gravity braiding. *Phys. Rev. D* **2020**, *101*, 024036.
19. Takahashi, K.; Motohashi, H. General Relativity solutions with stealth scalar hair in quadratic higher-order scalar-tensor theories. *arXiv* **2020**, arXiv:2004.03883.
20. Emparan, R.; Johnson, C.V.; Myers, R.C Surface terms as counterterms in the AdS-CFT correspondence. *Phys. Rev. D* **1999**, *60*, 104001.
21. Brihaye, Y.; Radu, E. Black hole solutions in d= 5 Chern-Simons gravity. *J. High Energy Phys.* **2013**, *2013*, 49.
22. Tchrakian, D.H. A remark on black holes of Chern–Simons gravities in 2n + 1 dimensions: n = 1, 2, 3. *Int. J. Mod. Phys. A* **2020**, *35*, 2050022.
23. Bekenstein, J.D. Black hole hair: 25-years after. In Proceedings of the Moscow 1996, 2nd International A.D. Sakharov Conference on Physics, Moscow, Russia, 20–24 May 1996; pp. 216–219.
24. Volkov, M.S.; Gal'tsov, D.V. Gravitating non-Abelian solitons and black holes with Yang–Mills fields. *Phys. Rep.* **1999**, *319*, 1–83.
25. Herdeiro, C.A.R.; Radu, E. Asymptotically flat black holes with scalar hair: A review. *Int. J. Mod. Phys. D* **2015**, *24*, 1542014.
26. Banados, M.; Henneaux, M.; Teitelboim, C.; Zanelli, J. Geometry of the 2+ 1 black hole. *Phys. Rev. D* **1993**, *48*, 1506; Erratum in **2013**, *88*, 069902.

MDPI

St. Alban-Anlage 66

4052 Basel

Switzerland

Tel. +41 61 683 77 34

Fax +41 61 302 89 18

www.mdpi.com

Symmetry Editorial Office

E-mail: symmetry@mdpi.com

www.mdpi.com/journal/symmetry